WHEATON PUBLIC LIBRARY

W9-AVK-487

Souper Tomatoes

Souper Tomatoes

Andrew F. Smith

RUTGERS UNIVERSITY PRESS
New Brunswick, New Jersey, and London

The material contained in this book is published solely because of its historical interest and is not intended as a source of recipes for the modern reader or of medical information or advice. Neither author nor publisher assumes any responsibility for the reader's application of the material contained herein.

Library of Congress Cataloging-in-Publication Data

Smith, Andrew F., 1946–
Souper tomatoes : the story of America's favorite food / Andrew F. Smith
p. cm.
Includes bibliographical references.
ISBN 0–8135–2752–X (alk. paper)
1. Cookery (Tomatoes) 2. Tomatoes. 3. Tomatoes—New Jersey.
4. Soups. I. Title.
TX803.T6S648 2000
641.6'5642—dc21 99-25680
CIP

British Cataloging-in-Publication data for this book is available from the British Library.

Manufactured in the United States of America

To my brother, Steve Smith, and
to friends who have encouraged me
to write about food history and other matters:
Frank Dantona, Fred Czarra, Karen Hess,
Joe Carlin, Midge Longley, and
Norman Schreiber

Contents

Acknowledgments

This story could not have been told without the assistance of many others. I wish to particularly thank Karen Hess, culinary historian, for her usual and unusual advice and comments about the first few chapters of this book, and Joe Carlin, Food Heritage Press, for his constant encouragement, his location of soup recipes and articles, and his review and comments on the manuscript. Special thanks to my wife, Tatiana Kling, and Michael Beiser for their editorial comments about the book and for his translations of a German book on soup. I also greatly appreciated the assistance of Robert Lewis, the editor-in-chief of the Middle English Dictionary at the University of Michigan, Ann Arbor; Bonnie Leppert, public affairs coordinator, Campbell Soup Company, Camden, New Jersey; John Smith and Mat Wrbican of the Andy Warhol Museum, Pittsburgh, Pennsylvania; Tim Hunt, exclusive agent for prints and photography, the Andy Warhol Foundation for the Visual Arts, New York; Richard Orzalli, director, Agricultural Operations, Campbell Soup Company, Sacramento, California; Fran DuVernois, former vice president, Global Operations, Campbell Soup Company, Camden, New Jersey; Jackie Finch, senior manager, Global Consumer Food Center, Campbell Soup Company, Camden, New Jersey; David Young, founder, Soup Collector Club, Wauconda, Illinois; Patti Campbell, founder, CC International, Ltd., Ligonier, Pennsylvania; Nancy Bailey, collector of Campbell Soup ephemera, Delmont, New Jersey; Robert Brower, El Sobriente, California; Gloria Stiens, Santa Maria, California; Mary Lou Dantona, Simi Valley, California; and Maria-Louise Sideroff, Ramsey, New Jersey. As usual, I have greatly appreciated the assistance of librarians at the New York Public Library; the New-York Historical Society; the Library of Congress in Washington, D.C.; Rutgers University libraries in New Brunswick, New Jersey; the American Antiquarian Society in Worcester, Massachusetts; the Culinary Archives and Museum at Johnson & Wales University, Providence, Rhode Island; and the British Library, London.

I would also like to thank the book dealers who greatly helped locate works difficult to find in libraries. These include Jan Longone, Food and Wine Library in Ann Arbor, Michigan; Janet Jarvits, Cook Books in Burbank, California; Peter Luke, Ephemera, Old and Rare Books in New Baltimore, New York; Carol Greenburg, Cornucopia in Brattleboro, Vermont; Nach Waxman, Kitchen Arts and Letters in New York; and Keith Crotz, The American Botanist in Chilicothe, Illinois.

While all comments and responses have been appreciated, not all have been accepted or incorporated; I accept all responsibility for errors that may appear in this work.

Souper Tomatoes

Introduction

Many food books and magazine articles proclaim that Robert Gibbon Johnson ate the first tomato in America on the courthouse steps of Salem, New Jersey. According to some accounts, a large crowd gathered to witness this momentous event dated at September 26, 1820. Believing that the tomato was poisonous, they expected Robert Gibbon Johnson to froth at the mouth and keel over. But, after eating a tomato, Johnson did not die. His brave act shocked the crowd and changed the course of American culinary history. Ever since, Americans have eaten tomatoes, or so the story goes.

What bothered me about this oft-told tale was that details changed from one rendition to another, but no primary sources were offered in support of the various versions. As I do not live far from Salem, straightening out the facts of the story appeared to be a manageable research task. What I assumed would be a quick trip, a little research, and a fast solution turned into many visits over months. My life as a culinary detective had just commenced. I collected various renditions of the story, checked local letters written during the early nineteenth century in Salem, reviewed diaries and regional histories, read Salem newspapers, and explored national gardening and agricultural periodicals published around 1820. Robert Gibbon Johnson was an influential member of the Salem community, and he was into agricultural innovation. He was also a historian: he wrote the first history of Salem and was instrumental in launching the New Jersey Historical Society. But I found no mention of the tomato-eating incident in Johnson's written works, nor in any other primary source. In fact, the first reference to this story I could locate was published in 1908, eighty-eight years after the purported event. This account simply stated that Johnson ate a tomato in 1820. It offered no supporting evidence but could well have been true. Johnson owned a book with a tomato recipe, so he was aware tomatoes could be eaten. My first conclusion was that the real story was not about Johnson eating a tomato: it was about how a statement evolved into a colorful local tale and ultimately achieved the stature of an honored national legend. I pieced together how each author embellished the story with more dramatic twists and submitted an article to *New Jersey History,* the magazine of the New Jersey Historical Society. My first article on culinary history was published.

While researching Johnson folklore, I became consumed with the question of who really did eat the first tomato in America and how the tomato reached culinary stardom in such a short time. Based on my investigations I published several academic articles and eventually a book, *The Tomato in America: Early History, Culture, and Cookery* (Columbia, South Carolina, 1994), chronicling the tomato's history to the Civil War. Immediately upon its completion, I began researching a sequel covering the historical period from the Civil War to the twentieth century. While doing so, I found that the tomato story diverged in three directions. One was the commercial production of tomato ketchup. Researching this strand led to the publication of *Pure Ketchup: A History of America's National Condiment* (Columbia, South Carolina, 1996). A second direction was the rapid conversion of the tomato fruit from a ribbed, hard-cored, and frequently hollow blob to the delicious, juicy, lip-smacking treat created at the end of the nineteenth century. The beginning of this story was chronicled in the introduction and appendix to *Livingston and the Tomato* (Columbus, Ohio, 1998). A third direction branched toward the tomato processing industry, the subject of this work.

This book employs a particular food product—tomato soup—as a lens to view the histories of soup, tomatoes, and canning. The first chapter covers the origin and conception of soup beginning in the prehistoric world and continuing through the nineteenth century. By this time, homemade soup was an important part of America's culinary scene. The second chapter examines the introduction and adoption of the tomato in Europe and America. The third chapter looks at the development of homemade tomato soup in America. The fourth chapter chronicles the rise of the canning industry, particularly in New Jersey. Tomato canning led directly to commercial tomato soup, which is discussed in the fifth chapter. The sixth chapter explores the revolutionary changes that have rocked the tomato soup industry during the past century. The final chapter inquires into tomato soup as an icon of American life and the current trends that affect the food.

When I began researching the canning industry, I had come full circle. Although Robert Gibbon Johnson did not eat the first tomato in America, New Jerseyites made major contributions to the tomato's history. The first tomatoes were canned in Jamesburg, New Jersey. When Union Army contracts stimulated the canning industry during the Civil War, New Jersey became a major tomato growing state. This growth expanded after the war, making South Jersey the top tomato growing area in the United States. Many tomatoes were marketed fresh to customers in metropolitan areas, but most went to the canneries that sprang up throughout the state. In 1869, a partnership was created that eventually became the Campbell Soup Company. It was not the first company to can soup, but it quickly became the dominant soup manufacturer in America. Its Camden plant was the world's largest canning factory. Campbell's Agricultural Research Division in Cinnaminson developed tomato varieties, including the J.T.D., named after the company's president, John T. Dorrance, the developer of Campbell's condensed tomato soup. Lyman G. Schermerhorn of the New Jersey Experiment Station at Rutgers University developed the Rutgers tomato in 1928, based on a Campbell researcher's cross of the J.T.D. with the Marglobe tomato variety. By the early 1950s the Rutgers tomato was used by over 70 percent of the tomato canners in the United States.

Immediately before and after World War II, tomato growing and canning were extremely important economic activities in South Jersey. But during the 1960s the tomato industry in New Jersey collapsed. Gone today are the canning factories that once dotted the South Jersey landscape. Gone are the vast fields of tomatoes blooming in the spring and fruiting in the fall. Gone are the aromas of cooking tomatoes during the harvest season. New Jersey's rivers no longer run red with tomato culls at canning time. The soup factory that once was the world's largest and a mainstay of Camden's economy has been demolished. Tomato growing is now mainly limited to providing fresh tomatoes for grocery stores and farmer's markets.

The significance of the tomato in Salem's history certainly contributed to the invention and continuation of the Robert Gibbon Johnson myth. But when New Jersey's tomato industry died, the Johnson story continued to thrive. This awareness helped me understand the making of the legend from a different perspective: the Johnson tale thrived in Salem because it fostered a remembrance and celebration of things past and pride in the community's history.

Despite major dislocations, the souper tomato story continues to unfold. The Campbell Soup Company retained its headquarters in Camden, and it remains the world's largest soup manufacturer. The tomato is still the queen of American vegetables; the manufacturing of tomato products is at an all-time high; soup is one of America's favorite foods; tomato soup is the fifth largest selling dry good in America and one of the most popular soups in the world. Without the contributions of New Jersey tomato growers, breeders, researchers, transporters, processors, marketers, and companies, souper tomatoes would not have achieved stardom.

While I am concerned with what tomato soup can tell us about larger social and historical issues, I remain fascinated by the tomato itself. As a child growing up in California, I consumed tomatoes in many guises. Like Andy Warhol, I was served tomato soup by my mother for lunch; along with sandwiches, it was a luncheon staple. Subsequently, I have continued to enjoy tomato soup in various forms and all its distant relatives. Its history is a juicy tale filled with unexpected twists and turns. It is action-packed, peopled with archaeologists and anthropologists, seedsmen and farmers, grocers and scientists, efficiency-conscious processors and managers, commercial artists and hard-hitting advertisers, and just plain old everyday consumers—all of whom have contributed to the transformation of tomato soup into one of America's favorite dishes.

▪ CHAPTER 1 ▪

Primordial Soup

Soups are diverse. In their simplest form, they contain liquid enriched with almost anything edible: fowl, flesh, fish, seafood, fruit, vegetables, cereals, milk, wine, salt, spices, seaweed, or even birds' nests. Their temperatures fluctuate from steamy to chilly to icy. Their consistencies range from thin to thick to chunky. Some are clear; others are creamy. They may contain a single, spiced ingredient; others blend multiple components so that none predominates; and still others maintain the integrity and individuality of solid ingredients. Some soups are served as appetizers for the meal, while others are the meal.

Despite this diversity, soups have common characteristics. By definition they have a predominance of liquids as opposed to solids and are served in a bowl or mug. They can be consumed with only a spoon. Still, soup's borders remain fluid. Where indeed are the boundaries between soups and sauces, stews, gravies, or other liquid dishes? The challenge of definition is partly due to the diverse linguistic roots of what is today called *soup*. Over the centuries, *potage*, *broth*, *consommé*, and *bouillon* have vied with the term *soup* for supremacy in the English language. To make this matter more confusing, the word *soup* has shifted in meaning over the years, as have the other related terms, making it difficult to determine what was actually meant at a particular time. Today's all-encompassing conception of soup is a relatively recent development: the word *soup* did not fully acquire its overarching status in English until very late in the eighteenth century.

Of course, consuming heated liquids in bowls or other containers preceded modern conceptions. Anthropologists maintain that our distant relatives domesticated fire at least a million years ago. This revolutionary discovery was initially employed for warmth and protection. The archaeological record does not pinpoint precisely when premodern humans first used fire for cooking, but abundant evidence of butchering and roasting appears about three quarters of a million years ago in Africa. Whenever the discovery was made, it was momentous. Roasting plant and animal foods permitted exploitation of a wider array of consumables. It made them more digestible, nutritious, and flavorful. Depending on the temperature and duration of exposure, roasting food killed harmful bacteria and parasites and neutralized some poisons. It also provided a means of preserving foods during times of plenty for subsequent periods of scarcity. Finally, roasting created new taste sensations, such as that of caramelized flesh.

Compared with roasting, boiling is a relatively recent discovery. The major reason for its late arrival on the culinary scene was the absence of easily available containers that were waterproof and heatproof. Reay Tannahill, author of *Food in History,* illustrated the cumbersome methods that would have been necessary for prehistoric humans to boil food. They dug large pits in the ground and lined them with flat, overlapping stones, which prevented seepage. Water was then poured or channeled into the pit. Hot stones from the campfire were pushed, pulled, or lugged into the water and added regularly to maintain its temperature. Meat or other food was then cooked in the hot water. As awkward and inefficient as this method appears, stone boiling was employed by isolated stone-age peoples well into modern times.[1]

Foods were also boiled in natural containers, such as mollusc or reptile shells in the Amazon, bamboo sections in Indonesia, large stone cooking pots in the ancient Tehuacán valley in Mexico, and leather containers and animal stomachs in Europe and North America. Of the latter technique, the ancient Greek historian Herodotus reported that, when Scythians did not have a cauldron, "they put all the flesh into an animal's paunch, mix water with it, and boil it like that over the bone-fire. The bones burn very well, and the paunch easily contains all the meat once it has been stripped off. In this way an ox, or any other sacrificial beast, is ingeniously made to boil itself." Similar methods were employed by Native Americans: the Lakota filled buffalo stomachs with meat and water and heated the stomach-containers over the fire.[2]

Stone boiling was also used to cook vegetables. In the American Southwest, E. W. Gifford observed a unique method of stone heating. Traveling Yavapai cooked food by placing it along with hot stones in scooped-out barrel cacti; the stones heated the cactus juice and thus the food.[3] In the eastern part of North America, the Chippewa, using two sticks, picked up hot stones and placed them into water-filled clay vessels to boil water; with a wooden ladle, the soup was transferred to clamshells, which served as soup bowls.[4] Before the French introduced iron kettles into New France, the Iroquois put water and meat into bark dishes, then threw a red-hot stone "into this fine soup." When one stone cooled, it was replaced with a hot one. The soup required constant stirring to prevent the stone from burning a hole through the basket. Fr. Paul LeJeune, a Jesuit living in New France in 1633, reported that Native Americans were fond of *sagamite*, a warm cornmeal gruel. As soon as indigenous groups became acquainted with European cooking techniques, *sagamite*'s meaning expanded "to signify all sorts of soups, broths, and similar things."[5]

Although pottery was invented ten thousand years ago, early clay pots were hardly suited for soupmaking, as they broke when exposed to heat. While Old World prehistoric piles of fire-cracked rocks suggest that hot stones were used for heating liquids, and historical evidence supports the belief that hot stones were dropped into pottery filled with food and water, it remained a cumbersome process. Widespread boiling probably did not begin until the invention of earthen-fired vessels about five thousand years ago. Laboratory testing of residues demonstrates that New World pottery was also heated over a fire and that souplike dishes were common in the pre-Columbian Americas.[6] Humans began making soup in earnest sometime between then and 300 BCE, when large copper and bronze kettles were manufactured for the express purpose of heating liquids.[7] Fireproof pottery and copper pots could withstand repeated heating on the outside and simultaneously retain liquid on the inside.

Boiling food greatly altered lifestyles in the prehistoric world. One French ethnologist, Germaine Tillion, equated soupmaking with civilization. She proclaimed that early soup cauldrons were extremely heavy and difficult to move. Soupmaking thus exemplified the radical transformation of the hunter-gatherer lifestyle to a sedentary one. Soup civilizations were characterized by particular diets, modes of living, and seasonal rhythms. In many places in the world, this pattern of soupmaking continued for thousands of years; in some places, it still survives.[8]

Boiling water proved to be an advantageous cooking technique compared to roasting with hot air over a fire. Because the water comes in contact with the entire surface of submerged foods, it easily and quickly imparts its energy to the food. In addition, boiling cooks food at a constant temperature, usually lower than frying, roasting, or baking; before the invention of thermometers, this was particularly significant. Boiling also permitted the fuller use of animal and plant products and expanded the range of foods that our ancestors consumed. Many animal parts, such as bones, could not be eaten even if roasted. Boiling extracted whatever nutritional value these previously unused parts possessed. Likewise, some plant parts were inedible, but became consumable only after boiling. For instance, acorns are edible only after boiling water removes the tannin.

Soaking food in clay pots softened the fibers of roots, leaves, seeds, and hard fruit, rendering them more digestible and nourishing. Like roasting, boiling foods created new taste sensations. When boiled, for instance, cereal grains released starch granules into the liquid and caused a sudden thickening; how thick depended on the amount and type of cereal grains. Heating several ingredients in a liquid over time caused each particular foodstuff to lose its individual taste and combine with the others to form a uniquely flavored mixture. The addition of animal products or oil-bearing seeds made the concoction richer, thicker, and more nutritious.[9]

ANTIQUARIAN PORRIDGE

Precisely when humans first consumed the liquid in the boiling pot is unclear. As most primitive peoples wasted nothing, it is extremely likely that the liquid was consumed along with the boiled contents as soon as it was technologically possible to scoop out the liquid. According to the earliest records, soup and porridge-type dishes were prepared by many ancient peoples.

From neolithic times, soup was consumed in the Mediterranean. Then, as now, the soup people ate reflected their social class.[10] In classical Greece and Rome, lentils, beans, and peas were converted into souplike gruels. According to culinary historian Andrew Dalby, lentil soup was typically consumed every day by workers and may well have been promoted by the philosophers, but it was "unlikely to be seen at rich men's feasts." In first-century Rome, Marcus Apicius, the culinary gastronome who compiled the only cookery manuscript that survives from the ancient Mediterranean, offered gourmet recipes for Barley Soup, Kale Soup, and Vegetable Soup.[11] Although no longer considered fashionable dishes, similar recipes are still found in modern Italian cookery.[12]

After the collapse of the Western Roman Empire, soup survived in the Byzantine Empire centered around Constantinople. With its fall to the Ottoman Turks in 1454, soups of Central Asian origin entered Europe's culinary repertoire. For many in Ottoman Turkey, bread and soup was the meal itself. Traditional soups included *tutmaç* with small meat-filled dumplings and "wedding soup" made with meat and flour, as well as yogurt, chicken noodle, and lentil soups. In the mid-sixteenth century, soup was a common dish. A Spanish slave who served as physician to a Turkish admiral, for instance, compared a Turkish soup composed of chicken and rice to an Italian minestrone. Unlike Western Europeans, the Turks did not limit soup consumption to a particular time of day or a specific course during a meal. Vegetables were used extensively in Turkish soups, and in later times the tomato was no exception.[13]

National cuisines in Western Europe emerged about the same time as Turkish culinary styles solidified. Surviving cookery manuscripts from the Middle Ages show a great deal of overlap in medieval recipes originating in England, France, Italy, and Catalonia. These manuscripts were intended for use by chefs of the nobility—often, specifically chefs of the king's household. Since they were written for chefs with extensive experience, specific directions and quantities were often omitted, making the author's intentions difficult to determine today. As most cooks were unable to read or write, the manuscripts were likely written and read by noblemen, who passed on their contents. Cooks learned their trade through apprenticeships, and few were honored with the title of chef. Those achieving this status often traveled from one nobleman's household to another, which may explain the overlap in recipes from manuscripts written in widely separated geographical areas. While similarities existed among foods eaten by Europe's nobility, the common people ate mainly what was produced locally. Often these foods, particularly cereals, were boiled and consumed in bowls. Today it would be more accurate to refer to these dishes as porridges or gruels rather than soups.

GERMANIC SOP

The Germanic invasions of Western Europe introduced two culinary terms that affected our modern-day conception of soup: the first was *broth* and the second was *sop*. Both Germanic root words survive in modern English, French, Italian, and other Western European languages. The term *broth* initially meant "that which has been brewed" and derived from the same root as the modern English word *brew*. From the earliest times, the word *broth* was also used to denote the liquid in which something was brewed. Broths were created by boiling and straining meats and vegetables. Not until the seventeenth century was broth generally associated with the liquid derived from boiling meat. By this time, the process for making broth was evidently important, as the proverb "Too many cooks spoil the broth" dates to this period.[14]

Initially, broths were served in bowls shared by more than one person at table. The liquid was usually sipped directly from the bowl. While broths were a basic foodstuff among the common people throughout Europe, upper classes considered the solids in the prepared dish to be the important ingredient: the liquid may have been intended as a sauce, to impart a particular flavor rather than to be consumed in its own right. The hot liquid also served to keep the solids warm. Sometimes the meat or vegetables used to create the broth were eaten separately; other times they were cut into smaller pieces and served in the broth. But when this happened, according to Terence Scully, culinary historian of the Middle Ages, "the diner normally had no desire to scoop out or drink any remaining liquid as we would a modern soup." Prior to the common adoption of table spoons and the invention of the fork, diners used their right hands to pull out chunks of meat or vegetables from shared bowls. For those more fastidious, knives speared the solids and conveyed them to the mouth.[15]

When fashion shifted in fourteenth-century Europe, spoons served as vehicles to transport the liquid from the bowl to the mouth. Then, as now, careful balancing was necessary to avoid spilling the liquid. The solution to the challenge of dining on liquid food was the sop. According to Robert Lewis, the editor-in-chief of the *Middle English Dictionary* at the University of Michigan, *sop* meant "a dish containing pieces of bread, usually toasted, on which a broth or syrup is poured." Middle English *soupe* was borrowed from an Old French noun with the same meaning, which came ultimately from a Germanic root—the same root that developed into Modern English *sop, sup,* and *supper.* The interesting linguistic point, notes Lewis, is that Middle English had two words from the same root that came into the language via different etymological routes: *soupe* borrowed from the Old French and *soppe* directly from Anglo-Saxon.[16] In France and Italy, *sop* continued to refer to the bread or toast. An interesting recipe in French and Italian manuscripts for a dish variously called *suppe dorate* or *soupys yn dorye* was actually an early version of what we would consider a type of French toast today. In the English language, the term *soup* evolved to describe the liquid in which the bread or toast was soaked.[17]

While soup continued to be consumed directly from the bowl and through sops right up to the nineteenth century, the spoon gradually became the dominant mode of conveyance, at least among European upper classes. This change may have been encouraged during the late sixteenth century when men and women started wearing large, stiff lace collars called ruffs. Those wearing ruffs around their necks could not easily drink soup from bowls without spilling, and hence fashion dictated the use of spoons. But early spoons were unsuitable for soup consumption. Since most spoons had fairly short stems, soup could not be safely negotiated past the ruffs without spilling. So spoons acquired longer handles and larger bowls, permitting more liquid to be transported to the mouth with less chance of dribbling the contents on the ruffs.[18]

RENAISSANCE SOUP

The Renaissance that transfigured the fine arts in Europe also unleashed a culinary revolution. Cookery in the Middle Ages largely reflected the availability of foods grown locally. As the spice trade revived during the Crusades, Arab cookery influenced the culinary arts, particularly in Spain and Italy. During the Renaissance, Italian and French cooks broke away from traditional styles, experimenting particularly with reviving the ancient Greek and Roman cookery.[19]

From the fifteenth century, several Italian cookery manuscripts survive. Among the more important is the work of Maestro Martino de Rossi, a celebrated chef of Cardinal Ludovico Trevisan. Written in the 1460s, Martino's manuscript reflects the major changes underway in Renaissance Italy. A copy of the manuscript fell into the hands of Bartolomeo Sacchi, who was appointed librarian of the Vatican. Using the pseudonym of Platina, Sacchi added sections to the manuscript and published it under the title *De honesta voluptate et valetudine* or, translated, *On Right Pleasure and Good Health* (Venice, 1470). It was the first published cookbook, and it was filled with recipes for consommés, broths, potages, and soups.[20]

Italian Renaissance cookery reached a high point a century later with the publication of Bartolomeo Scappi's *Opera dell'arte del cucinare* (Venice, 1570), a monumental culinary achievement rivaling the artistic creations of Leonardo da Vinci and Michelangelo. It included dozens of recipes for diverse soups and potages, which were generally composed of hot broth or liquid poured over hard bread (sop) or another foodstuff. Recipes for potages can be found in many cookery manuscripts from different European regions, predating the publication of cookbooks. Some potages were thick stews, while thinner ones would be classed as soups today. The vestigal notion of a potage survives into the modern day in the form of French onion soup or the use of croutons in soups. Scappi described the preparation of *minestra,* which literally translated as "thin soup" but which also often denoted the first course of a meal. Adding the suffix *-one,* meaning "largeness," resulted in *minestrone,* which meant a hearty, chunky vegetable soup probably originating in Genoa. Scappi also offered recipes for *brodo* (broth) and for *suppa* or *zuppa,* which, like the potage, was served on or with toasted bread.[21] Scappi's massive work laid the foundation for Italian cookery, and many of his dishes survived into modern times.

One chapter in Scappi's work focused on recipes for the sick featuring the proverbial chicken soup. The idea that certain soups were curative derived from the principles of humoral medicine. According to the doctrine of humors, each food had specifically prescribed qualities or attributes. The action of the stomach was analogous to a furnace that cooked food. The problem was that various foods had different cooking times in the stomach. Some foods had qualities that eased digestion, while others tended to hinder it. Illness often resulted from stoppage or delay in the digestive process. As thin foods were believed to be easier to digest, broths were recommended for the sick, a recommendation that survives today long after the principles of humoral medicine have disappeared.[22]

Modern social scientists have offered other reasons for the association of soup with health. For instance Vance Packard proclaimed in his *Hidden Persuaders*: "Besides being a good food, stimulating to the appetite and easily assimilated into the bloodstream, soup is unconsciously associated with man's deepest need for nourishment and reassurance. It takes us back to our earliest sensations of warmth, protection, and feeding. Its deepest roots may lie in prenatal sensations of being surrounded by the amniotic fluid in our mother's womb. No wonder people like soup and prefer it hot and in large quantities. They associate it with the basic source of life, strength and well-being." As interesting as such theories are, other societies do not have these psychological associations with soup. It is more likely that our Western beliefs are remnants of now forgotten humoral theories.[23]

There are other remnants. Humoral theories posited that easily digested foods should be consumed first during a meal and heavier foods toward the end. The cook's job was not just to prepare the meal but also to sequence foods properly. The health of his patron depended upon decisions about the composition of each "prepared dish[;] so too it depended upon the sequence in which he decided to set out his various preparations on his patron's dining board," states culinary historian Terence Scully.[24] This medieval European concept created the notion that soups were appropriate for the first course of the meal.

ENGLISH SOUP

Recipes for souplike dishes appeared in several cookery manuscripts from medieval England. For instance, *The Forme of Cury,* compiled in the late fourteenth century by Richard II's chefs, offered a recipe for *Sowpes in galyntyne* made from sugar, salt, and a sauce composed of flour, vinegar, and several other ingredients. The manuscript also contains a recipe for cabbage potage.[25] In a fifteenth-century manuscript, *soupes dorre* was composed of strained and ground almonds, saffron, sugar, and several other ingredients boiled together and served as a potage. Another recipe for *soupes chamberlyne* consisted of wine, ginger, sugar, and other ingredients. Yet another manuscript featured *Lyode Soppes* with boiled milk, salt, and sugar and *Oyle Soppys* made of onions, saffron, pepper, sugar, and salt. All of the above concoctions were poured over toasted bread or sops.[26]

While broths and sops were commonly consumed in England, the English associated the term *soup* with the French during the sixteenth and seventeenth centuries. Some Englishmen believed that French soups were not quite honest. An anonymously written *Satyr against the French* (London, 1691) mentioned French dishes "which few Mankind know beside: With Soops and Fricasies, Ragou's, Pottage."[27] At about the same time,

British poet and playwright John Gay associated soup with snails when he wrote: "Spongy morells in strong ragouts are found, And in the *soup* are the slimy snail in drown'd."[28] About forty years later Jonathan Swift wrote contemptuously:

> She sent her priest in wooden shoes
> From haughty *Gaul* to make ragoos:
> Instead of wholesome bread and cheese,
> To dress their soops and fricassees.[29]

Despite these disparagements, soups, potages, and broths were significant components of British cookery. In 1593 Andrew Broode wrote: "Potage is not so moch used in al Crystendom as it is used in Englande." Broode recommended potages of all types.[30] Gervase Markham's *English Hus-wife* (London, 1615) presented several recipes for potages, some with sippets or sops used to soak up the juice.[31] In *The Closet of the Eminently Learned Sir Kenelme Digby, KT., Opened* (London, 1669), the "Chancellour to the Queen Mother" featured many potages and broths in his cookbook, such as *Potage de santé* and just "plain" English potages. Included as well were recipes for broths made in the Portuguese way and others composed of barley and spinach. One broth was titled "Broth for sick and convalescent Persons."[32] Soups were also incorporated into *The Accomplisht Ladys Delight* (London, 1675), attributed to Hannah Woolley, one of the first female English-language cookbook writers. Woolley's Rich Barley Soup was thickened in the medieval manner with breadcrumbs. Woolley inevitably presented a recipe for chicken soup in the medicinal section.[33]

The most extensive seventeenth-century English treatment of potages, broths, and soups is found in Robert May's *Accomplished Cook* (London, 1660). May received his chef's training in Paris but served his apprenticeship in London. Subsequently, he was employed in thirteen different households of minor English nobility. He was knowledgeable about food trends in Europe: while he did not fully break with the prevalent food customs of the Middle Ages, he did introduce new culinary concepts to England. Among his English potages May embraced recipes for French bisques and Italian "Brodos." His "Soops" called for savory ingredients, such as spinach, carrots, artichokes, potatoes, skirrets, and parsnips. He offered broth recipes for the sick. Like previous writers, May maintained the medieval custom of pouring the liquids over toasted bread. Of the two hundred pages in the book, potages and soups consume almost 20 percent. May's diagrams for table settings featured a large potage dish in the middle of the table.[34]

In the eighteenth century, English interest in soup increased. E. Smith's *Compleat Housewife*, first published in 1727, featured recipes for Gravy

Soup, White Soup, and Oyster Soup. Smith also published two for crawfish or lobster soup, two for strong broths, two for Asparagus Soup, and four for Pea Soup.[35] Charles Carter's *Complete Practical Cook* (London, 1730) advertised "Directions to make all Sorts of excellent *Pottages* and *Soups*" in its title. On the first page, Carter asserted: "The chief Source of this Part of Cookery is the Strong Broth Pot; for a good Stock of strong Broth well made, and good Gravies well drawn off, are very principal Ingredients in the composing of all Made-Dishes of boil'd Meats." Carter then offered dozens of recipes for potages, bisques, broths, and soups, including ones for Cabbage Soup, Cucumber Soup, and Pocket Soup (a solid mass converted to soup by adding water).[36]

Mid-eighteenth-century British cookbooks brimmed with soup and broth recipes. Hannah Glasse's popular *Art of Cookery* (London, 1747) featured thirteen recipes in a chapter titled "Soops and Broths" and fifteen more in the chapter "For a Fast-Dinner, A Number of good Dishes, which you may make use of for a Table at any other Time." Eight more recipes for broth and soup were scattered throughout the rest of the work, including five recommended for the sick. Her Beef or Mutton Broth recipe was *"for very weak People, who take but little Nourishment."*[37] Glasse also published an early recipe for Scotch Broth, which was a compound of meat stock, traditionally mutton, thickened with barley and peas or other vegetables. In Scotland it was called barley broth, and many recipes for it had been published previously. Samuel Johnson, the British lexicographer, critic, and conversationalist, ate bowlfuls of this broth with gusto, although he otherwise slighted the diet of the Scots. Johnson later defined soup as a "Strong decoction of flesh for table."[38] Subsequent editions of Glasse's *Art of Cookery* incorporated a dozen more soup and broth recipes. The 1768 edition included one "To make Chouder, a Sea Dish," which produced a stew-like dish.[39] It featured pork as well as biscuits but lacked the potatoes, milk, and tomatoes common to nineteenth-century chowders.[40]

Glasse's rivals were Susannah Carter's *Frugal Housewife* (London, 1765), which counted nineteen soup recipes, and Elizabeth Raffald's *Experienced English Housekeeper* (Manchester, 1769), which presented eleven pages of soup recipes, such as Portable Soup for Travellers, Transparent Soup, Hare Soup, Vermicelli Soup, and Ox-Cheek Soop. Other soups were based on almonds, onions, and peas.[41]

By the early nineteenth century, soup recipes were widely published in almost all British cookery books. Alexander Hunter's *Culina Famulatrix Medicinae* (London, 1802) incorporated sixty-seven soup and broth recipes, one-fourth of all recipes in his cookbook. In addition to previously mentioned soups, Hunter featured ones based on carrots, giblets, macaroni, partridges, hares, and sheep's heads.[42] Richard Dolby's *Cook's*

Dictionary (London, 1830) borrowed heavily from previously published cookbooks, but it featured dozens of recipes for soups, consommés, potages, and broths.[43]

FRENCH SOUP

The French made significant strides in systematizing their classical cookery during the late seventeenth century. Its basic building blocks were stocks composed of beef, pork, poultry, fish, or vegetables. From stocks, sauces, gravies, and soups were generated. Classic French soups were categorized into an elaborate system of which broths, bisques, consommés, and purées were a part. Simultaneous with the codification of classical French cookery was the glorification of regional cuisines, such as those of southern France that created bouillabaisse and those of Brittany that produced chowders. This tension between regional and traditional cookery gave French cuisine a dynamic that brought it to greatness in the eighteenth century.

In the late eighteenth century, another French culinary invention was the modern restaurant. While taverns and inns were present throughout Western Europe, they mainly catered to travelers. The fare served at these establishments was intended as accompaniment for the alcoholic beverages. Travelers took what they could get, and rarely was there a choice of foods. A "M. Boulanger" is usually credited with establishing the first restaurant in 1765. The word *restaurant* had existed previously, referring to a broth or stock with restorative powers. Over the door of Boulanger's soup shop were the words "*Boulanger débite des Restaurants divin,*" which translated meant "Boulanger pours out heavenly refreshments." Along with the sign was a motto: "*Venite ad me omnes qui stomacho laboris, ego vos restaurabo,*" which translated as "Come to me all ye with a hungry stomach, and I will restore you." Boulanger's soups were not particularly novel: his *le consommé restaurant* was a refreshing meat broth composed of stock from partridges, mutton, and veal boiled in a *bain-marie* (double-boiler) for twelve hours, after which the "restaurant" or restorative was pressed through a cloth.

Boulanger placed a table or two in his shop at which customers could enjoy their fare. He wanted to sell other foods, but due to convoluted French laws, he could not do so. However, as business was not brisk, he decided to push his luck and sell sheep's feet. The Parisian meat *traiteurs*, or guilds, challenged Boulanger in court. Boulanger won his case before the Paris Parlement. Due to the court case, his fame spread far and wide, or so the story goes. Boulanger's connection with the word *restaurant* was mentioned in the 1771 *Dictionnaire de Trévoux*. Six years later, *L'Almanach Dauphin* defined *Restaurateurs* as those who mastered the art of making meat broths (called *Restaurants* or *Bouillons de Prince*). Their new establishments that served these dishes were called *restaurants*.[44]

ECONOMICAL SOUP

Another important milestone in the soup saga was the invention of economical soup by Benjamin Thompson, who was born in Massachusetts in 1753. Like an estimated third of those living in the British North American colonies, Thompson was a Loyalist. At the commencement of the American Revolution, he sided with the British. When the British evacuated Boston, Thompson went first to Canada, then to England. He served in the British War Department and returned to America to command the Queen's Horse Dragoons against the revolutionaries. After the war was lost, he went back to England and worked on a number of scientific experiments. When he received an offer of employment in Munich, the British government, which wanted better relations with Bavaria, knighted Thompson. While in Munich, Thompson at the request of the Bavarian government organized a system of assisting the poor by building working-class housing and improving their diet. To feed them economically, he developed a thick soup composed of pearl barley, peas, carrots, slices of wheat bread, wine vinegar, salt, and water. This was called "poor man's" or "economical" soup in Germany (*Sparsuppe*). It was intended to be the cheapest and most nutritious food that prevented malnutrition and preserved health.[45]

By 1800 Thompson's economical soup had spread throughout Western Europe, as had his fame.[46] His good works earned him the title of Count Rumford in Bavaria. But he did not rest after this success. He was instrumental in inventing and introducing a new type of stove and other cooking equipment. As important as his other achievements were, his "poor man's soup" was a lasting creation that became even more significant as the nineteenth century unfolded.

Throughout Europe, soup kitchens serving Rumford's soup were established. For many of Europe's destitute, it was their only nourishment. To avoid famine in 1812, for instance, Napoleon dished out about two million free servings every day in Central Europe. Rumford's economical soup was featured in many cookbooks of the nineteenth century.[47] Later analysis by J. C. Drummond and Anne Wilbraham indicates that Rumford's formula was based on economics rather than nutritional requirements. His recipe contained less than one thousand calories—half the number necessary daily for working men and women.[48] Still, Rumford had worked at a time before there was extensive knowledge of nutrition, and economical soup helped many of Europe's poor to survive.

Others picked up where Rumford left off. In 1845 a bad grain harvest struck Europe, resulting in an escalation of wheat and bread prices. When the blight destroyed the potato crop, soup kitchens were established throughout Europe to avert starvation. The Irish were particularly dependent on the potato, and they suffered grievously when their potato crop continued to fail. Starvation was widespread. To make matters worse, the English Corn Laws prevented the importation of grain supplies from other countries. Simultaneously, locally grown grain was shipped out of Ireland to meet contractual agreements with English merchants. Famine resulted, and an estimated two million Irish died or emigrated.

When famine broke out, Alexis Soyer started a long correspondence with the London *Times* intended to raise funds for soup kitchens in London and Dublin. A Frenchman by birth, Soyer lived his adult life in England, where he served as the chef of London's famous Reform Club. Prior to the famine, the soup kitchens distributed about forty gallons of soup to about two hundred people daily. Soyer proposed plans for creating larger and more efficient soup kitchens to overcome widespread famine. In January 1847 the British government invited Soyer to construct a prototype soup kitchen. Instead of erecting it in London, Soyer opened it three months later in Dublin, where he believed it was needed most. It was a temporary shelter built of canvas and wood, housing a steam boiler with a three-hundred-gallon capacity and an oven for baking bread. The soup was cooked for about an hour and was then poured into thousand-gallon *bain-maries*. When it was ready, a bell rang and one hundred people were admitted through the front door. Soup bowls were given to each person and were filled. As soon as everyone was seated, grace was said and soup consumption commenced. Precisely six minutes later, the bell rang and the group filed out the back door, while another group was admitted.[49]

Soyer's kitchen fed eight thousand people per day during the first few months. At the height of the famine, it fed twenty-six thousand people daily. The soup kitchen was so politically successful that the British government bought it and handed over its operation to a community group. Based on his experiences, Soyer published six economical soup recipes in the London *Times,* hoping that "they may prove advantageous by giving a change in food, which acts as generously on the digestive organs as a change of air does on the convalescent, and likewise to prevent the rise in price of any particular articles."[50]

Soyer was the hero of the hour, and ballads about him were sung by an admiring public. One titled "Ode to Soyer" lauded his actions and predicted that his name would be handed down through the generations as a true "Regenerator." Soyer liked the thought. Building on the proposed accolade, Soyer published the *Poor Man's Regenerator; or, Charitable Cookery*, which sold for sixpence. For every copy sold, he donated one penny to charity.[51]

Soyer's work encouraged other chefs to work with the poor. Queen Victoria's chief cook, Charles Elmé Francatelli, published *A Plain Cookery Book* (London, 1852), which is thought to be the first cookbook written for the working class. Francatelli, British born but of Italian parentage, had received his culinary training in France. Apparently, he had some strange ideas about working-class cookery: he included over thirty gourmet soup recipes, such as one for bouillabaisse. His book did, however, feature several economical soup recipes for the poor.[52] Not to be outdone, Soyer responded with *Shilling Cookery for the People* (London, 1855), featuring thirty-nine soup recipes. In this cookbook, Soyer reported that many Englishmen did not favor soup because recipes in most cookbooks were so complicated and expensive that most people could not "afford either the money, time or attention, to prepare it." He simplified recipes so that soups could be quickly made and were nutritious, wholesome, and economical.[53]

AMERICAN SOUP

Some historians have incorrectly concluded that soup was not at all a part of the diet in colonial North America. According to Richard Hooker, Americans neglected soups until almost the end of the nineteenth century.[54] To support this claim, Hooker cited the accounts of foreign observers. Gottlieb Mittelberger, a German who visited America in the mid-eighteenth century, reported that the American colonists knew "little or nothing of soup eating."[55] A French visitor, reported Hooker, described American dinners as having broth but no soup. Hooker quoted a Swiss immigrant who prepared his own food on the American ship *Xenophon*: "Americans want nothing to do with soup, and the sailors and officers almost die laughing when they see us prepare our soup every day."[56] Another foreign visitor concluded in 1869 that American dinners were "usually wanting in soup."[57]

While some Americans may have eschewed soup, most evidence indicates that soup was an important dietary component in the young country, at least among the upper and lower classes corresponding to those who ate it in Europe. Even before cookery books were published in America, newspapers, magazines, and travel accounts mentioned broth and soup, as well as recording recipes. For instance, the *Gentleman's Magazine* offered "A Recipe for a Soup for Tristram Shandy."[58] William Byrd, who helped establish the borders between Virginia and North Carolina, consumed venison, turkey, rice, and barley soups. His pocket or portable soup was "so strong, that two or three Drams, dissolv'd in boiling Water, with a little Salt, will make half a Pint of good Broth." According to Byrd, "One Pound of this Cookery wou'd keep a man in good heart above a Month, and is not only nourishing, but likewise very wholesome. Particularly it is good against Fluxes, which Woodsmen are very liable to, by lying too near the moist Ground, and guzzling too much cold Water. But as it will be on us'd now and then, in times of Scarcity, when Game is wanting, two Pounds of it will be enough for a Journey of Six Months." Byrd recommended that "this Broth will be still more heartening if you thicken every Mess with half a Spoonful of Rockahominy, which is nothing but Indian Corn parched without burning, and reduced to Powder."[59]

Portable soups, immersed in water when wanted, were common in Britain and America during the eighteenth century. They were particularly resorted to by travelers and seafarers and were sold commercially. In 1775 Captain James Cook took portable soup, which he had found valuable for sick sailors, with him on his voyage around the world. Subsequently, other British ships employed portable soup on long voyages with varying degrees of success.[60]

William Parks, a colonial printer in Williamsburg, Virginia, published the first cookbook in America in 1742. It was based on the fifth edition of E. Smith's *The Compleat Housewife.* While Parks did not add recipes to the cookbook, he excluded those recipes he thought were inappropriate for life in America. Parks included recipes for Soop Sante, Pease Soop, Craw Fish Soop, Brooth, Soop with Teel, Green Peas Soop, and several bisques.[61] Susannah Carter's *Frugal Housewife,* published in America beginning in 1772, contained an entire soup chapter presenting nineteen soups. Neither Hannah Glasse's *Art of Cookery* nor Elizabeth Raffald's *Experienced English Housekeeper* was printed in America until the early nineteenth century, but colonists brought copies from England, and both were popular in colonial America.[62]

While the works by Smith, Carter, Raffald, and Glasse contained many soup recipes, they did not necessarily reflect culinary practice in America. However, three surviving colonial cookery manuscripts attest to the importance of potage, broth, and soup in America—at least among the upper class. From Boston, *Mrs. Gardiner's Family Receipts from 1763* featured directions for making soups and broths, as well as recipes for Ox Cheek Soup, Peas Soup, Lent or Fast Day Peas Soup, White Soup, Queen's Soup, Rice White Soup, Onion Soup, Soup Meagre, Portable Soup, Mutton Broth, and Scots Barley Broth. In this manuscript, the soup recipes account for more than 10 percent of the entire work. All Gardiner's soup recipes were copied from the cookbooks of Glasse, Raffald, and Carter. They are noteworthy not for their originality but for the fact that British colonists thought that soup recipes were important enough to copy for potential family use. Mrs. Gardiner's husband, Dr. Silvester Gardiner, was an outspoken Loyalist, as were her children and their families. While she died before the American Revolution began, most of her family along with other Loyalists, such as the previously mentioned Benjamin Thompson, left Boston on "Evacuation Day." The Continental Army under the command of George Washington commandeered drugs Dr. Gardiner left behind, prompting Gardiner to refer to their commander as "that thief Washington."[63]

The second work was based on a late seventeenth-century British manuscript that had been in the family of Martha Custis, who married George Washington. This *Booke of Cookery* included recipes "To make French Pottage" and "To make French Broth." Washington indeed served soup at his dinners, so the soups in this manuscript may have been served to those dining with him at Mount Vernon.[64]

The third manuscript was Harriott Pinckney Horry's *Receipt Book* of 1770, which is thought to have derived mainly from the cookery manuscript of her mother, who had been born to English parents in the West Indies. The manuscript included two references to soups, one noted in a recipe to dress a calf's head, and the other in a recipe for preserving tomatoes for making soup. Horry's father had been the British lieutenant governor of Antigua. When the American Revolution broke out, her husband fought on the colonial side until Charleston fell, when he "took protection" from the British and sailed for England, leaving his wife behind to handle their plantation. He returned after the peace treaty was signed but died disgraced in 1785. Harriott Pinckney Horry's two brothers fought bravely on the American side during the war, so when George Washington toured South Carolina in 1791, he breakfasted and dined with Harriott Horry.[65]

While American culinary traditions were based on English cookery, as the manuscripts discussed above attest, many other national and cultural groups contributed to the cuisine of the new nation. For instance, German immigrants had influenced culinary matters in America since the late seventeenth century. Pennsylvania-Germans liked soups and were particularly famous for those based on potatoes.[66] When Jacques Pierre Brissot de Warville ate at the home of a wealthy Philadelphia Quaker, the meal included two soups.[67] Subsequent German-language cookbooks published in America featured recipes for soups based on chicken, mutton, veal, beef, calf's head, rice, apples, and huckleberries. According to food historian William Woys Weaver, the Pennsylvania-Germans took their commitment to soups seriously. Soup was a "symbol of community, of religious fellowship, and even communion." In two-course meals, soup was the first; in one-pot meals, soup was the only dish.[68]

The most influential culinary tradition in America after that of the English was French cookery. As a result of the French Revolution, many noblemen and their families and entourages fled France. Chefs were among these refugees. Some found positions in homes of the wealthy in England and other countries. Others opened restaurants. The French gastronome Jean-Anthelme Brillat-Savarin, who wisely left France for America at the height of the Reign of Terror, visited a café-tavern in New York where turtle soup was served for breakfast. While in America, Brillat-Savarin also visited Jean Baptiste Gilbert Payplat dis Julien. In 1794 Julien opened a public eating house in Boston called Julien's Restorator. This was his English translation of the French word *restaurant.* Julien was famous for his soups among gourmands, while the novelty of his cuisine attracted customers. Julien was nicknamed the "Prince of Soups."[69] He is credited with creating Julien Soup, a composition of vegetables in long narrow strings. Julien specialized in making turtle soup—the king of soups—with few spices. Julien stated that he had not created his Restorator to "deprive gentlemen of the felicity of dining with their families by overcharging their stomachs. The ingredients of this soup render it such that its effects are 'as if increase of appetite should grow by what it fed on.'" Julien died in 1805, but his widow continued his restaurant for ten years. She, in turn, turned it over to yet another French immigrant, Frederic Rouillard.[70]

Almost simultaneous with the French Revolution was the slave uprising in what is today Haiti. Many French refugees, along with their slaves and servants, immigrated to New York, Boston, Charleston, Philadelphia, and New Orleans as the terrors of the slave rebellion peaked. They brought their culinary traditions with them, and some also launched restaurants.

Perhaps due to this double French influence, soup was adopted by American upper classes. It was often the first course at fashionable dinner parties.[71] Neither was soup neglected by American cookbook writers. While the first edition of Amelia Simmons's *American Cookery* (Hartford, 1796) did not contain soup recipes, it did note that parsley was "good for soup." The second edition of this work, published in the same year in Albany, included recipes for soup made of a beef's back, veal, and lamb's head and pluck, as well as a recipe for chowder.[72] Mary Randolph's *Virginia Housewife* (Washington, 1824) featured sixteen soup recipes, such as ones for asparagus, beef, gravy, veal, bouilli, oyster, and barley. There were two for peas, more for hare or rabbit, fowl, catfish, onion, and turtle, and, finally, one for mock turtle soup using calf's head. Her directions for making turtle soup combined turtle flesh, beef, bacon, onions, sweet herbs, pepper, and salt. If a rich soup was desired, butter and flour were folded in. She recommended that soup be seasoned with wine, ketchup, spice, cayenne, and curry powder.[73]

N.K.M. Lee's *Cook's Own Book,* first published in 1832, copied the encyclopedic approach and the recipes of the previously mentioned British author Richard Dolby. *The Cook's Own Book* was essentially a compendium of simplified recipes compiled from diverse British and American sources. It included eighty-seven recipes for soups, consommés, and broths. Some were based on asparagus, beef, mutton, beet root, calf's head, carrot, celery, crawfish, cress, cucumber, eel, giblets, gourds, game, hare, herbs, lobster, macaroni, ox head, ox heel, oxtail, peas, pigeons, spinach, venison, vermicelli, and barley. Lee also furnished such unusual recipes as Lorrain Soup, Moor-fowl Soup, Mulligatawny, Curry Soup, and Cocky-Leeky Soup—a soup of Scottish derivation made with fowl and leeks.[74]

No pre–Civil War cookbook author, with the exception of Mary Randolph, made a greater contribution to American cookery than did Philadelphia-born Eliza Leslie. Leslie was well aware of French strides in improving culinary practices. In her *Directions for Cookery* she exclaimed that it was "not true that French cooks have the art of producing *excellent* soups from cold scraps. There is much *bad* soup to be found in France, at inferior houses; but *good* French cooks are not, as is generally supposed, really in the practice of concocting any dishes out of the refuse of the table." She advised against thickening soup with flour, which spoiled both the soup's appearance and taste. Her recipes included soups based on beef, mutton, cabbage, noodles, veal, macaroni, vermicelli, milk, venison, hares, rabbits, oxtails, peas, lima beans, asparagus, chicken, catfish, eels, lobsters, oysters, clams, and chowder. She included unusual recipes for Water Souchy, Meg Merrilies' Soup, and "Mullagatawny Soup as Made in India." Leslie provided a recipe for Mock Turtle or Calf's Head Soup, but she omitted "a receipt for *real* turtle soup, as when that very expensive, complicated, and difficult dish is prepared in a private family, it is advisable to hire a first-rate cook for the express purpose." She recommended going to a turtle-soup house and purchasing it rather than making it in the home.[75]

Not all cookbook writers considered soup to be healthful. Josepha Buell Hale was deeply opinionated about soup and how it should be served. Her *Good Housekeeper* (Boston, 1839) cited the experiments conducted by Connecticut surgeon William Beaumont and reported in his *Experiments and Observations on the Gastric Juice* (Plattsburgh, 1833). Beaumont, who examined the effect of gastric fluids on different kinds of food, had concluded that soups were "among the most indigestible." Based on Beaumont's studies, Hale asserted that dyspeptics and those suffering from bilious disorders or "troubled with heart-burn and indigestion, would be injured by eating soups often." But she considered soup generally beneficial for children, those not in good health, or those engaged in hard labor, especially if "a good share of rice and other vegetables be in the liquid, or considerable bread eaten with it." Hale urged serving soup at the beginning of the dinner, as otherwise diners "would be inclined to take too much solid food." Hale also believed in the importance of separating the blood from the meat, which "should be conscientiously attended to by the Christians as abstaining from pork is by the Jew." After the liquor was thus purified, cooks could then add vegetables—rice, carrots, cabbage, onions, or potatoes.

Contrary to Leslie's advice, Hale recommended thickening soups with flour. Hale also suggested adding crackers or toasted or hard bread a short time before the soup was served, but not "those libels on civilized cookery, called *dumplings!* One might about as well eat, with the hope of digesting, a brick from the ruins of Babylon, as one of the hard, heavy masses of boiled dough which usually pass under this name." Finally, Hale urged cooling the soup before consuming it, as "hot liquors greatly injure the teeth; and also by passing immediately into the blood and thus circulating through the whole system, cause an unnatural glow and perspiration which often predisposes to colds, and is weakening to delicate constitutions. When we are well, and wish to continue so, it is best never to take food or drinks warmer than milk." After these stern warnings, Hale offered recipes for Mock Turtle Soup, Currie Soup, Veal Soup, Beef or Mutton Soup, White Soup, Pigeon Soup, Vegetable Soup, and Rice Soup.[76]

Almost every other writer concluded that soup was a healthy addition to table. Recipes for soupmaking appeared in all general cookbooks, which gave proportionately greater space to soups as the century progressed.[77] Despite all the coverage, not every American knew how to make soup. The first known American cooking pamphlet focused solely on soups, *Soups and Soup Making* (Chicago, 1882), was written by Emma Ewing. She believed that soup was "convenient, economic and healthful." As an article of diet, it ranked second in importance only to bread, proclaimed Ewing. Soupmaking was "justly entitled to a prominent place in the science of cooking." Scientifically prepared, soup was "easier of digestion than almost any other article of diet." However, she warned that it "must not be a weak, sloppy, characterless compound, nor a crude, greasy, inharmonious hodge-podge. The defects of unsavory, unpalatable, indigestible soups may be concealed, but can not be removed by the excessive use of salt, pepper and other spices and condiments." Ewing believed that soup "must be skillfully prepared, so as to please the eye and gratify the palate."[78]

Popular cookbook writer Mary Virginia Terhune, who wrote under the pseudonym of Marion Harlan, published an article entitled "Soup Making" in the *Home-Maker* magazine in 1888. She reported that "the slovenly or indifferent housewife" produced "wishy-washy stuff" that tantalized "a healthy appetite only to fill the gastronomic soul with worse than naught." Others produced soup with "no substance *to* it!" Terhune believed that soup should be the "introduction to the ceremony of dining—the overture to the stately opera." She proclaimed that the French never omitted it: their preliminary course was soup, "light and varied in flavor and appearance." But many Americans were "ignorant of the value of *consommé* or French bouillon, for daily consumption. Boiled gently from paste, tapioca, sage, rice, barley, farina, or left clear and poured upon neatly-poached eggs in the tureen." Terhune concluded that soup was "nutritious, stimulating, and easily digested."[79]

Class distinctions continued to be reflected in soup consumption in America. Among the upper class, soup was associated with refinement: "The higher one rises in the social circle, the more nearly soup approximates the necessaries of life," reported an observer. Despite this positive image among the upper class, soup was a "neglected and much-abused article" that was cold-shouldered by "perhaps one-half of the homes of what may be called middle-class families." Among the lower class, soup was unknown "except as sick-room broth among the really poor." Terhune explained that the omission of soup arose "partly from the disinclination to increase the care of the servantless mistress by adding a course to the family dinner that involves the need of changing and cleaning an extra set of plates." But she also believed that ignorance of soupmaking skills was another cause. "To make good soup," claimed Terhune, "one must have patience and judgment, and take the time required for the various processes of manufacture." She believed that soup should be consumed every day, for it was "wholesome, economical, and, when properly made, delicious."[80]

Soup had not achieved the prominence that it deserved, according to other observers. Elizabeth Robins Pennell, a Pennsylvanian by birth who lived much of her life in Europe, was quite appreciative of soup, but in 1896 she asked:

> Equally desirable in illness and in health, during one's journeys abroad and one's days at home, why is it then that soup has never yet been praised and glorified as it should be? How is it that its greatness has inspired neither ode nor epic; that it has been left to a parody—clever, to be sure, but cleverness alone is not tribute sufficient—in a child's book to sing its perfections. It should be extolled, and it has been vilified; insults have been heaped upon it; ingratitude from man has been its portion. The soup tureen is as poetic as the loving cup; why should it suggest but the baldest prose to its most ardent worshipers?[81]

Pennell, who claimed to have amassed one of the largest collections of cookbooks in private hands, lambasted soupmaking in England and

America but highly praised it in France, particularly the soup served to travelers, who greatly appreciated "the delicate, strong, refreshing, inspiriting *bouillon,* served at every *buffet."* French soup helped "one to forget fatigue and dust and cinders" and gave "one a new heart to face the long night and the longer miles. Soup sustains and nourishes; and, better still, it had its own aesthetic value, for it was the one perfect dish for the place and purpose." A particularly important ingredient, according to Pennell, was the tomato: "In soup, thin or clear, the tomato knows no rival."[82]

Whether Americans consumed as much soup as the French is unknown. However, soup was an important part of most Americans' diet by the end of the nineteenth century. By the beginning of the twentieth century, the United States saw an ever sharper rise in soup consumption. As a dish that was both economical and nutritious, soup was prepared for men in prisons and the military and offered without charge at lunch in saloons. Because it was simultaneously inexpensive and considered a gourmet dish, it was also just right for the rapidly expanding American middle class.

Using the pseudonym of Olive Green, Myrtle Reed stressed soup's economic side. Consuming soup also was a fine test of table manners. "To sip silently from a spoon, with apparent pleasure, a sizzling hot liquid which one does not like," reported Reed, was as "fine a tribute to one's host as that paid by a foreign diplomat." Yet she firmly believed that so many soups abounded that one could be found to please almost everyone: "When the flavor of a wild beast is unpleasing, a tame one may be relished, and *vice versa.*"[83] Reed's *One Thousand Simple Soups* (New York, 1907) was the largest single collection of such recipes published anywhere up to that time. Many of her soup recipes featured tomatoes.

■ CHAPTER 2 ■

The Original Tomato

When the first Europeans arrived in the New World, no tomatoes grew in what is today the United States. How and when tomatoes were introduced there has been a matter of conjecture for over a century. The most common explanations are based on folklore, such as those associated with Robert Gibbon Johnson eating the first American tomato in 1820 on the courthouse steps of Salem, New Jersey. Little primary source evidence for these theories has been uncovered, and the current renditions featured in popular accounts are unlikely to be accurate.[1]

What is known about the tomato's introduction into the United States presents a confusing picture. The plant originated along the coastal highlands of western South America, but there is no evidence that indigenous peoples in South America ate tomatoes before the arrival of the Spanish. In pre-Columbian times the undomesticated tomato migrated by unknown means to Central America, where it was domesticated by Mesoamerican peoples. When the Europeans arrived during the early sixteenth century, tomatoes were only consumed in a rather narrow geographical area of Central America and Mexico, with Mexico City at its northern boundary. This lack of widespread diffusion in pre-Columbian times has led many observers to conclude that tomatoes were a late addition to the culinary repertoire of Mesoamerica.

The Spanish first encountered the tomato after their conquest of Mexico in 1519–1521, yet few references to tomatoes have been located in Spanish colonial documents. In 1529 Bernardino de Sahagún, a Franciscan, came from Spain to Mexico, where he spent the remainder of his life. He learned Nahuatl, and under his direction, Aztec priests compiled materials in their native language. Sahagún then translated the oral Nahua words into close approximations to the Spanish language. Sahagún was the first European to make written note of "tomates." According to Sahagún, Aztec lords combined them with chile peppers and ground squash seeds and consumed them mainly as a condiment served on turkey, venison, lobster, and fish.[2] This combination was subsequently called *salsa* by Alonso de Molina in 1571.[3]

THE TOMATO DIASPORA

The tomato arrived in Spain and Italy shortly after Hernan Cortés's conquest of Mexico. Tomatoes were first mentioned in an Italian herbal published in 1544. This source recommended that tomatoes should be cooked just like eggplants and mushrooms. Likewise, other accounts from Spain indicate that tomatoes were consumed in Seville by 1600.[4] By the late seventeenth century, the first known tomato recipe published in a cookbook appeared in *Lo scalco alla moderna* (Naples, 1692) by Antonio Latini. While the work was published in Naples, some recipes were of Spanish origin.[5] Within a hundred years, tomatoes were well established in Italian cookery—at least in the southern regions. Vincenzo Corrado's cookbook, *Il cuoco galante opera mecconica* (Naples, 1786), featured dozens of tomato recipes with several specifically for tomato soup.[6]

Tomatoes were commonly used throughout the Moslem Mediterranean. Lancelot Addison's *Account of West Barbary*, written in 1671, reported, "Besides the sallad ordinary in other Countries, they have one sort rarely to be met with in Europe which they call by a word sounding Spanish Tomate's. This grows in the common Fields, and when ripe is plucked and eaten with oil; it is pleasant but apt to cloy." Not only were tomatoes used in salads in North Africa, they were also soup ingredients. Speaking of Algiers and Tunisia during the 1720s, Thomas Shaw remembered seeing "tomatas," which were used as "a relish to their soups and ragouts."[7]

The tomato was similarly used in Spain, Portugal, and Italy. British botanist and agriculturist Charles Bryant reported in 1783 that the Portuguese and Spanish used it "in almost all their soups and sauces, and . . . deemed [it] cooling and nutritive."[8] British gardener John Claudius Loudon noted that "near Rome and Naples, whole fields are covered with it, and scarcely a dinner is served up in which it does not in some way or other form a part."[9]

England was one of the last Western European nations to adopt tomatoes. The first known references to tomato cultivation in England were written by John Gerard, a barber-surgeon and herbalist. His *Herball* (London, 1597) was largely copied from the works of other herbalists. In one of the original sections of the herbal, Gerard claimed that his "golden apples" or "apples of love" came from Spain and Italy. However, Gerard considered "the whole Plant" to be "of ranke and stinking savour." He classified the tomato as "colde, yet not fully so colde as Mandrake, after the opinion of

Dodonaeus: but in my judgement it is very colde, yea perhaps in the highest degree of coldeness." The fruit was corrupt, which he left to every man's censure. Yet, under the category of "Virtues," Gerard reported that "in Spaine and those hot regions they use to eat the apples prepared and boiled with pepper, salt and oile: but they yeelde very little nourishment to the bodie, and the same nought and corupt. Likewise they doe eat the apples with oile, vinegar and pepper mixed together for sauce to their meat, even as we in these cold countries doe mustard."[10]

Gerard's apparent schizophrenic response—that the tomato was corrupt in England but consumable in Spain and Italy—can only be understood within the context of the previously mentioned humoral system of medicine, which defined the medical properties of plants based on four related qualities: hot and cold, wet and dry. Degrees were assigned to each quality. The highest degree of each quality was considered poisonous or toxic. Gerard likely judged the tomato "cold" simply because it was filled with moisture or juice. The humoral system also dictated that the effects of foods were related to the season and outside temperature. If one partook of a cold food during winter, illness might result. But it was perfectly all right for people in hot countries to consume cold foods, as Gerard depicted in Spain and Italy. Gerard's assignment of tomatoes to the highest degree of coldness was in agreement with the assessments of other herbalists from "cold" countries, such as the Flemish herbalist Rembert Dodoens, but not with those from "hot" countries where tomato cookery thrived.[11]

THE SOUP TOMATO

British attitudes toward the tomato began to change during the early eighteenth century as travelers ate tomatoes outside England and returned to tell the tale. British herbalist William Salmon reported that tomatoes in Spain were "boiled in Vinegar, with Pepper and Salt, and Sewed up with Oil and Juice of Lemons; Likewise they eat them raw, with oil, Vinegar and Pepper, for Sauce to their Meat as we do Cucumbers." Salmon quickly pointed out that tomatoes "yield not much nourishment," but he readily admitted that they "please and cool or quench the Heat and Thirst of hot Stomachs."[12] In 1742 the Quaker merchant Peter Collinson reported that tomatoes were "very much used In Italy to putt when Ripe into their Brooths & soops giving it a pretty Tart Tast." Although he had not tried them, he reported that an English lady with more courage than he thought that tomatoes gave "an Agreeable Tartness & Relish to them & she Likes it much" in her soup. One or two tomatoes were put in at "a Times, the boiling Breaks them & then they are Diffused through the whole."[13]

British gardeners and seedsmen grew tomatoes as ornaments, but some were interested in their culinary uses. Philip Miller, in his *Gardeners Dictionary* (London, 1748), offered the typical British refrain about tomatoes. However, the 1754 edition of this same work reported that tomatoes were "now much used in England" in soups, although he added that there still were "persons who think them not wholesome." In a later edition of this work, Miller opined that the tomato gave "an agreeable Acid to the Soup."[14]

Miller was not the first to publish a report that some Englishmen consumed tomatoes in soups. In 1753 John Hill reported that tomatoes were eaten "by the Jew-families in England as we do cucumbers, with oil, vinegar, or else stewed in soops."[15] Other English observers offered similar comments about the Jewish-tomato connection. In 1492 the Jews had been expelled from Spain. Most moved to the eastern Mediterranean and Eastern Europe. Only *conversos* or *morranos*—Jews who outwardly converted to Catholicism—remained in Spain. Shortly after Columbus's exploration of the Caribbean, many *morranos* immigrated to the New World, where they more easily escaped suspicion of practicing Jewish traditions or performing non-Catholic religious ceremonies. Many Jews in the New World engaged in trade. George Stedman, while traveling in Dutch Surinam in the 1770s, stated, "I found plenty of tomate, which being produced in many British gardens I will not attempt to describe; but only observe that the Jews are particularly fond of it, and stew it with butchers meat instead of onions."[16]

As England increased its global political and economic power, Jews familiar with New World foods established communities in London and other English cities. These communities also brought new foods to England. By the mid-eighteenth century, the tomato was an accepted food among Jews. In the nineteenth century, the Jewish-tomato connection continued. English horticulturalist Henry Phillips announced that the tomato had "long been used by the wealthy Jewish families in this countries; and within these last few years it has come into great use with all our best cooks."[17] When Judith Montefiore published the first known English-language Jewish cookbook, *The Jewish Manual* (London, 1846), she included one recipe for tomato sauce and two for tomato soup.[18]

During the late eighteenth century, many British gardening books described tomato culture and reported the use of tomatoes in soups.[19] The first known English-language recipe using tomatoes was published in a supplement to Hannah Glasse's *Art of Cookery Made Plain and Easy* (London, 1758). The recipe was titled "To dress Haddocks after the Spanish Way," so presumably it originated in Spain. A similar recipe was revised for publication in Richard Briggs's *New Art of Cookery* (London, 1788). Briggs's cookbook was in turn reprinted in Philadelphia three years later, becoming the first American work to publish a tomato recipe.[20] The 1797 edition of *Encyclopedia Britannica* reported that the tomato was "in daily use; being either boiled in soups or broths, or served up boiled as garnishes to flesh-meats."[21]

At the beginning of the nineteenth century, Alexander Hunter, a Scottish physician who was convinced that no one could "be a good physician who has not a competent knowledge of cookery," published a cookbook under the imposing title *Culina Famulatrix Medicinae; or, Receipts in Modern Cookery* (London, 1806). Hunter had studied medicine in France and had acquired many of his recipes from continental European sources. According to him, tomatoes had a pleasant acid taste and were much used by the Spaniards and Portuguese in making soups. They made "a charming sauce for all kinds of meat, whether hot or cold." Perhaps as important, Hunter included a recipe for Mock Tomata Sauce, made from sharp-tasting apples with turmeric for coloring. This at least demonstrated that by this date tomatoes were important enough for cooks to create their simulated appearance in a dish when not available.[22] Hunter's recipes were published regularly by other cookbook authors for over fifty years.

British gardeners grew tomatoes extensively in some regions. A.F.M. Willich claimed that the tomato was "greatly esteemed at the table of the epicure: it is either used in soups or broths, to which it imparts an agreeable acid taste; or it is boiled and served up as a garnish to dishes of animal food."[23] In 1819 British agriculturist William Cobbett observed that great quantities of tomatoes were sold in London, where they were employed to thicken stews and soups.[24] In the same year British cookbook author Maria Eliza Rundell assured her readers that tomatoes were "used in soups, sauces and to serve as little dishes at table, at any part of a dinner."[25] By 1831 Henry Phillips announced that "Love-apples" were "to be seen in great abundance at all our vegetable markets," but he did not find that they were "used by the middle or lower classes of English families, who have yet to learn the art of improving their dishes with vegetables." Phillips mentioned specifically that tomatoes when added to soup imparted "an acid of a most agreeable flavour."[26]

THE AMERICAN TOMATO

Although British gardeners and seedsmen brought tomatoes into what is today the United States, they were probably not the first to do so. The Spanish had likely cultivated them previously in their colonies and settlements in North America: Florida, California, and the American Southwest. As English settlers occupied territories previously controlled or influenced by Spain, such as South Carolina, Georgia, and Florida, they were exposed to tomatoes. The previously mentioned British herbalist William Salmon visited South Carolina in the 1680s. He noted that English colonists grew tomatoes and that they could be used for many medicinal cures. Some American colonists ate tomatoes as early as the mid-eighteenth century.[27] Tomatoes were also consumed in British colonies in the Caribbean, where many Americans had contact with them. In 1774 Edward Long reported that in Jamaica tomatoes were converted into a gravy and "used in soups and sauces and impart a grateful flavour."[28]

During the late eighteenth and early nineteenth centuries, British gardening books reprinted in America incorporated references to tomatoes. Charles Varlo, a farmer and agricultural writer in Scotland and Ireland, ventured to America in 1785. He traveled throughout New Jersey, eastern Pennsylvania, Maryland, and Virginia (where, of course, he dined with George Washington) and returned to England. While in Philadelphia, he published an edition of his *New System of Husbandry* that mentioned tomatoes. In 1799 an American edition of Charles Marshall's *Introduction to Gardening* was published in Boston. This gardening book mentioned red, yellow, white, and cherry-fruited tomato varieties: red and green varieties were good for pickles, while red varieties, claimed Marshall, strengthened soups. In 1803 an American reprint of the British work *Gleanings from the Most Celebrated Books on Husbandry, Gardening and Rural Affairs* noted two varieties of tomatoes, both of which were prescribed for medicine and also used in sauces, pickling, and soups.[29] Although published in America, these references reflected British usage of the tomato, not necessarily American.

Only one colonial cookery manuscript is known to have contained a tomato recipe. Its author, Harriott Pinckney Horry, copied a collection of recipes, including one titled "To Keep Tomatoos for Winter Use."[30] Other sources indicate that tomato cookery was well established in southern states by the mid-eighteenth century. From the southern states, tomato culture slowly spread up the Atlantic coast and into rural areas. Tomatoes were grown in Virginia by the 1780s. They were grown in Philadelphia during the 1790s.[31] Seedsmen sold tomato seeds in Philadelphia by 1800 and in New York and Baltimore shortly thereafter. Bernard M'Mahon's *American Gardener's Calendar,* published in 1806, stated that the tomato was esteemed for culinary purposes and was "much cultivated for its fruit, in soups and sauces, to which it imparts an agreeable acid flavour."[32] In 1804 James Mease of Philadelphia reported that the tomato was "greatly esteemed at table: it is either used in soups or broths, to which it imparts an agreeable taste."[33] Likewise, New York seedsman Michael Floy sold "Tomatoes or Love Apples for Soup."[34]

By the early nineteenth century, tomato recipes appeared in manuscripts and cookbooks. Several cookery manuscripts contained recipes for tomato ketchup and other tomato recipes.[35] The first known tomato recipe published in a cookbook authored by an American appeared in *The Universal Receipt Book* (New York, 1814), attributed to Richard Alsop.[36] Tomato recipes were commonly published in cookbooks by the 1820s.

THE FRANCO-AMERICAN TOMATO

While English practices dominated early American cookery, French traditions also were influential. Tomato seeds were sent from France to the United States. After Jefferson became the U.S. ambassador to France in the 1780s, he sent tomato seeds back to friends in what is today West Virginia.[37] During the 1790s tomato seeds from France were sent to Philadelphia's natural history museum, founded by Charles Willson Peale.[38] This flow of seeds from France continued for over a century, and some of the best seeds in America derived from French seeds. As has been discussed, French immigrants and refugees introduced tomato cookery into America during the French Revolution.[39]

The French influence also filtered into America through cookbooks. Louis Eustache Ude, the "ci-devant cook to Louis XVI" who had removed to England during the French Revolution, wrote *The French Cook* (London, 1813), which demonstrated the elegant cooking that was the mark of every fashionable gentleman; it was first published in America fifteen years later. It contained over fifty soup, potage, and related recipes featuring consommés and bouillons. Ude, of course, included extensive directions for making the fashionable turtle soup.[40]

Many early American restaurants featured French cuisine, and gourmet soups became the rage of America's upper class. Subsequent books of French cookery published in the United States kept Americans abreast of the latest soupmaking developments and highlighted the tomato as an ingredient in many recipes.[41] Louis Eustache Audot's *La cuisinière de la compagne et de la ville* was translated into English and published in America as *French Domestic Cookery* (Philadelphia, 1846). It featured a recipe for gazpacho, which Audot identified as "a favourite dish with the Andalusians."[42]

THE MEDICINAL TOMATO

Tomatoes may have been grown in New Jersey before the American Revolution. An article written in 1862 reported a conversation years previous in which a woman said that her father cultivated them before 1776 in New Jersey and "that they were held in great esteem by the family for that purpose, and that her mother prepared a fine catsup from them." The original cultivator was a Loyalist who immigrated to Nova Scotia during the American Revolution.[43] The earliest primary source pinpointing the tomato in New Jersey was George Perot Macculloch's farm journal, in which he mentioned the planting of tomatoes from 1829 onward in Morristown. Several secondary sources agree that tomatoes arrived during the 1820s in different parts of New Jersey. James Mapes, editor of the *Working Farmer* published in Newark, New Jersey, reported that tomatoes had been "long grown in our gardens as an ornamental plant, under the name of Love Apple, before being used at all as a culinary vegetable." About 1827 or 1828 Mapes occasionally heard of tomatoes being "eaten in French or Spanish families, but seldom if ever by others."[44] Charles W. Casper, an early South Jersey canner, reported that tomatoes were "brought to Salem, New Jersey in 1829 by some ladies from Philadelphia, nieces of a former citizen of Salem named I. Z. Coffie." The tomatoes were sown "in pots and the plants transplanted in the ground. They were a curiosity in those days being small, round and red."[45] Another observer stated that tomatoes had been introduced into New Jersey gardens only during the late 1820s. But by 1830 another observer reported that tomatoes received "much attention" and had become "a great favorite at every table."[46]

The latter comment was premature. Tomatoes did not become common until the late 1830s. This shift was largely due to two individuals, John Cook Bennett and Archibald Miles. In the fall of 1834, Bennett, a professor in a medical school in Ohio, lectured to his students about the positive effects of ingesting tomatoes, which he said cured numerous illnesses. In July 1835 Bennett publicly announced these astounding discoveries in an Ohio newspaper. The news article flashed across the nation. Bennett's claims promptly surfaced in New Jersey. The *Salem Union* reported that tomatoes successfully treated diarrhea, violent bilious attacks, and dyspepsia (indigestion) and prevented cholera. Bennett urged all citizens to eat tomatoes as they were "the most healthy article of the Materia Alimentary."[47] On August 11 Bennett's claims were published in Elizabeth's *New-Jersey Journal* and Woodbury's *Constitution, and Farmers' and Mechanics' Advertiser*. The *Constitution* followed this up by reprinting an article from the *Gettysburg Star* with the earthshaking claim that a large tomato (1 pound, 12 ounces) had been discovered. The editor snidely remarked that this was "not as large as ours."[48] In Morristown, the *Morris County Whig* published Bennett's claims on August 12. The editor commented that there was no vegetable that could be "cultivated with less care and trouble, and none which will, if properly treated, yield in greater abundance," and that Bennett's claims were "another inducement for their free use, besides the strong relish which most palates soon acquire for this delicious vegetable."[49] Newark's *Sentinel of Freedom* printed them on August 18 and the *Paterson Intelligencer* on the following day.[50] Other New Jersey newspapers continued to reprint these claims for months and years.[51]

Archibald Miles read Bennett's comments about the medicinal effects of the tomato and decided to produce tomato pills. In July 1837 he launched "Dr. Miles' Compound Extract of Tomato." According to his advertisements, these pills cured dyspepsia (indigestion), headaches, intermittent and remittent fevers, ill-conditioned ulcers, hepatic and other cuticular affections, hepatitis, complaints of the liver, local congestions, bilious colic, all complaints requiring an aperient, and glandular and other concealed affections of the skin. For children, this medicine "cured Summer Complaint, Whooping Cough, Measles," and many other diseases. While Miles did not claim that his tomato pills cured all diseases, he did maintain that there was "no *one* medicine that will succeed more frequently." By early 1838 his pills were advertised in almost every part of the country, including New Jersey.[52]

The success of the campaign on behalf of "Dr. Miles' Compound Extract of Tomato" was immediate. An estimated one hundred thousand people reportedly consumed tomato pills throughout the United States and the Carribean. Success, however, brought competition and scrutiny from the medical community. By 1840 "every variety of pill and panacea" reportedly was composed of "extract of Tomato." Despite their early success, tomato pills were unmasked as a hoax, and the national market collapsed in the early 1840s.[53]

Most observers admitted that the tomato was a healthy esculent even if they disagreed with the wild assertions of the pill vendors. Tomatoes of all types were considered medicinal. Tomato soup was so identified by Lettice Bryan, whose recipe in her *Kentucky Housewife* (Cincinnati, 1841) was undoubtedly influenced by the popularity of the tomato pill, as it was located in the section titled "Preparations for the Sick."[54] Bryan was not alone in identifying tomato soup as a healthy dish. Medical professionals believed the same. For instance, Russell Thatcher Trall's *New Hydropathic Cook-book* (New York, 1854) strongly asserted the positive medicinal influence of tomato soup. After presenting his recipe, Trall exclaimed, "This is one of the most agreeable and wholesome of the 'fancy dishes.' "[55]

Despite their growing acceptability, tomatoes were still eschewed by many Americans, particularly in New England. Some may have thought them to be poisonous. However, more significant, if mundane, reasons emerged to explain popular revulsion. Some found that the reeking smell of the tomato plant made them nauseous; some believed that this odor was a warning against consumption of its fruit. Others did not like the tomato's appearance. One gardener stated that the first time he saw tomatoes, they "appeared so disgusting that I thought I must be very hungry before I should be induced to taste them." The look of the tomato was so disagreeable that many people supposed that it would "never receive a permanent place in our list of culinary vegetables." Likewise, many people found the tomato repugnant on first tasting it. J. B. Garber posited that "hardly two persons in a hundred, on first tasting it, thought that they would ever be induced to taste that *sour trash* a second time." A writer in the *Genesee Farmer* stated that the tomato was offensive in whatever shape it was offered. A writer in the *Horticulturist* found tomatoes to be "odious and repulsive smelling berries." Even these hostile judgments paled in comparison to those of Joseph T. Buckingham, editor of the *Boston Courier,* who said that they were "the mere fungus of an offensive plant, which one cannot touch without an immediate application of soap and water with an infusion of *eau de cologne,* to sweeten the hand—tomatoes, the twin-brothers to soured and putrescent potato-balls—deliver us, O ye caterers of luxuries, ye gods and goddesses of the science of cookery! deliver us from tomatoes."[56]

After the Civil War, Dio Lewis, a physician at Harvard's medical school and one of the founders of the physical culture movement, reported that when he was a youth, his mother advised him not to handle tomatoes. As a medical professional, Lewis collected plenty of evidence indicating that his mother had been right. At his public lectures, members of the audience testified to ills brought on by consuming tomatoes. Some complained of stomach conditions caused by eating raw tomatoes; others, that their piles were the result of excessive consumption of tomatoes. Still others suffered "from tender and bleeding gums, from 'teeth set on edge,' and quite a number from loose teeth, produced by eating tomatoes."[57]

As the tomato became more popular, Lewis became alarmed. At a lecture he asked, "Didn't you know that eating bright red tomatoes caused cancer?" Pennsylvania physician Dr. John Hylton took this point up, claiming that tomato cells were identical to cancer cells under the microscope. He further asserted that "there was much cancer where tomatoes were eaten." Some Americans were deeply concerned about these allegations and wrote to the editors of agricultural magazines and newspapers asking about these statements. The editor of the *American Agriculturist* claimed they were extremely unlikely to be true. A housekeeper from Newark, New Jersey, wrote in a letter to the *New York Tribune*: "My enjoyment of the tomato has of late been much disturbed by many unpleasant rumors with regard to its pernicious effects, such as tumors, cancers etc. Some say that only those are hurtful which are attacked by the tomato worm, and still others, among whom I am told are some excellent physicians, consider the free use of them under any circumstances injurious. Will not the doctors come forward and tell us what they think of the matter? If tomatoes are productive of such serious results surely we ought all to know it and banish them from our tables."[58]

Two months later Professor L. P. Arnold of Rochester, New York, declared tomatoes to be healthful. He wrote: "Tomatoes contain neither cancers nor cancer-producing matter." Tomatoes did have some defects: they were not, "like milk, a perfect diet of themselves, and besides, like most other articles of food, they contain some obnoxious qualities." But Arnold reassured readers that tomatoes should not be thrown out of the diet because of this: "Nature has provided us with such efficient organs that obnoxious matter in our food, if in moderate amount, is readily cast out, and the body is protected against any material injury." Still, Arnold cautioned "feeble persons" and those with "peculiar constitutions" about eating tomatoes.[59]

In Michigan a Dr. Homer Hitchcock decided to systematically survey physicians, asking for their evaluation of Hylton's claims. All responded negatively. One physician stated that "tomatoes will make cancer as soon as a red thread around the neck will stop the nosebleed." Despite the fact that this charge was denied by many physicians and other prominent individuals, it lingered for years. The matter came up again in 1888. The editor of the *Country Gentleman* reported that there was not the "slightest possible basis for this belief." In the following year A. O. Black of Vermont wrote in the *American Agriculturist*: "The idea that eating tomatoes can in any manner cause cancer is utterly baseless. It is remarkable that such a superstition ever started, and still more so that it has any currency since its total falsity has been shown up repeatedly." A writer in *Good Housekeeping* magazine in 1892 reported that "the pretended discovery was made, and widely circulated through the press, that the use of tomatoes was productive of cancerous affections, and the people were warned against exposing themselves to the danger." In fact, the correspondent asserted, the opposite was true: "It has been shown to the satisfaction of all thinking men and women that the tomato is one of the most healthful products of the vegetable world. It is decidedly helpful in cases of biliary disturbance, and its timely use often saves the necessity for a medical course."[60]

Like most American controversies, this one spread to the United Kingdom. In 1894 a writer wanted to know who originated the prevalent idea that tomatoes were a dangerous food: "So many inquiries came to the London Cancer Hospital concerning the relation of the tomato to that disease, that the physicians in charge found it necessary to make a public statement to this effect, 'that tomatoes neither predispose to, nor excite cancer formation, and that they are not injurious to those suffering from this disease, but, on the contrary, are a very wholesome article of diet, particularly so if cooked.' "[61]

Despite widespread acceptance, some physicians continued to attack the healthfulness of the tomato. Dr. W. T. English of Western University of Pennsylvania reported at the 1896 annual meeting of the American Medical Association that the tomato had a poisonous effect upon one-half to one-fourth of those who consumed it. It caused, English reported, "irregularity and inequality of heart rhythm and force, and the effect upon those with previous heart disease has been deplorable." He declared also that ingesting tomatoes caused high blood pressure, sighing respiration, skipped heartbeats, cold sweats, and disturbed vision. English maintained that general depression was "a common sequence to their employment in one of the alms houses near Pittsburgh" and in several hospitals. "The supplies for the military should never include this vegetable," he declared, for "it can not be for the physical good of the men." The tomato was "treacherous" and could "not be relied upon as a food." Those called upon to endure loss of sleep or suffer mental strain, he said, such as "the student, the speaker, the salesman, will find them worse than useless."[62] At the same meeting, two other physicians expressed the opposite opinion: they believed that the symptoms English described "were due to fermentation in the stomach from use of tomatoes picked before ripe."[63]

Most medical professionals praised the tomato in general and tomato soup in particular. *Hall's Journal of Health* recommended tomato soup as a healthy dish. Horatio Wood, a leading Philadelphia physician, reported that tomato soup was a "very elegant and cheap soup, suitable to many cases of invalids." He recommended combining three quarts of strained tomatoes with three pints of milk thickened with flour, along with butter and salt.[64] The Seventh-Day Adventist surgeon John Harvey Kellogg, the superintendent of the sanatarium in Battle Creek, Michigan, believed in the healthfulness of the tomato. His wife, Ella Kellogg, published recipes for Tomato Soup and Tomato Cream Soup, as well as many other soup recipes with tomatoes as an ingredient, such as Tomato and Rice Soup, Pea and Tomato Soup, Tomato and Macaroni Soup, Tomato and Okra Soup, Tomato Soup with Vermicelli, Celery and Tomato Soup, Nut and Tomato Bisque Soup, and Lentil and Tomato Soup.[65] A friend of the Kelloggs', Almeda Lambert, featured numerous soup recipes with tomatoes and nuts in her *Guide for Nut Cookery* (Battle Creek, 1899). One recipe employed Nutmeatose, a commercial health food composed of peanut butter, water, gluten, and salt.[66]

Concerns about the tomato's healthful qualities lingered well into the twentieth century. A spirited defense of the tomato was launched by the Vegetable Growers Association in 1924. Articles attributed to the organization proclaimed that physicians believed that tomatoes were "the richest of all vegetables in the natural health acids which keep our stomachs and intestines in condition" and served as "a gentle, natural stimulant which helps wash away the poisons which cause disease and contaminate our systems." Physicians prescribed tomatoes for diabetics and those suffering from Bright's disease. Medical authorities even reported that canned tomatoes were to be preferred to fresh. At the Children's Memorial Hospital in Chicago, physicians fed "tomatoes and tomato juice to little suffering waifs and starving infants (not milk, mind you, but tomatoes) and the babies are crowing in their cribs, bright eyed, rosy cheeked and getting well." Dr. Hugo Friedstein of Chicago said that "the vitamin content alone of tomatoes is accomplishing the undreamed of in feeding infants and children, and doing marvelous things in cleansing the system." Also the tomato possessed three "wonderful" acids: malic acid, also found in apples; citric acid, also found in lemons, limes, and other fruits; and phosphoric acid, used in the treatment of nervous disorders and other disturbances of health. According to the article, tomatoes had been used for centuries to relieve pain. Dr. Arnold Lorand of Carlesbad, Austria, reported that this healthful acidity was what gave the characteristic "tomato flavor." There was nothing "like it for the invalid and the convalescent; there is nothing better as a prime 'pick-me-up' and reviver for athletes, and there is no better food obtainable for everyday use for rich and poor alike." Lorand also reported that tomatoes relieved constipation. Canned tomatoes were "said to be the swiftest, surest and most certain natural remedy for obesity." Dr. D. G. Wagner, formerly a captain in the U.S. Medical Corps, agreed with these assessments and added that physicians believed tomatoes were the "richest of all vegetables in natural health acids, which keep the stomach and intestines in condition," and served as diuretics, thus "washing away poisons that cause disease." London's Dr. P. J. Cammidge ranked "the tomato first of all vegetables and fruits as a food treatment." At Johns Hopkins Hospital, physicians specified tomatoes as part of the diet for diabetes. Several physicians suggested that consuming tomatoes was better than injecting insulin.[67]

Despite this strong defense, the controversy lingered. Cookbook authors Anna Lindlahr and Henry Lindlahr felt compelled to state: "**Tomatoes do not make cancer**, but help to cure it. Most of our cancer patients, at one time or another while undergoing treatment, usually during the healing crisis, develop strong appetite for tomatoes, and we always encourage them to satisfy this craving to the fullest extent."[68] As late as the mid-1930s Della Lutes reported that cancer "was a word with which we were somewhat familiar, since some of our more conservative neighbors would not touch tomatoes on account of how Ob Hutchens, who ate them, had developed this dread disease."[69] It is ironic that the tomato, once thought by some to cause cancer, is now considered a potential cancer fighter.[70]

TOMATO COOKERY IN NEW JERSEY

New Jersey's first published cookbook, the anonymously written *Economical Cookery* (Newark, 1839), featured recipes for preserving tomatoes, tomato sauce, and ketchup, as well as two for tomato pickles.[71] The second cookery book published in New Jersey was a reprint of Phineas Thornton's *Southern Gardener and Receipt Book*. Thornton had been born in New Jersey but moved to Camden, South Carolina, where he became a successful merchant. Thornton's *Southern Gardener* (Camden, South Carolina, 1840) was intended to be "an everyday kind of book, a farmers' almanac and guide, and how-to manual for the independent rural Southern family." It featured five tomato recipes. The second edition was published five years later in Newark, New Jersey, and included two additional tomato recipes.[72]

In spite of the paucity of early New Jersey cookbooks, tomato recipes popped up in agricultural and gardening magazines during this period. For instance, "L" of Milton, New Jersey, submitted a recipe "To Preserve Tomatoes for Winter Use" in the Albany *Cultivator* in 1847. The tomatoes preserved in this manner were "as fresh in flavor and appearance as on the day they were taken from the vines." Late in the season, tomatoes were "*thoroughly* stewed, put into large mouthed glass bottles, such as are used for pickles, sealed tight when cold, and kept in a basement room." The writer believed that the success of this method was the result of their thorough cooking and recommended that the tomatoes "be stewed until the watery parts are evaporated, and the pulp changed to a crimson color."[73]

During the 1840s and 1850s, tomato cultivation flourished throughout the nation, particularly around large cities such as Philadelphia. New Jersey tomato grower Edmund Morris began farming in 1855 on a ten-acre farm in southern New Jersey. He shipped tomatoes to markets in both Philadelphia and New York. On one acre, he grew 3,760 tomato plants, which produced five hundred bushels of fruit and netted him $190. Morris noted that older hands at the business did much better, occasionally generating as much as four hundred dollars per acre. He remarked that good pickers gathered from sixty to seventy bushels a day. He estimated that the expense of cultivating tomatoes ran about sixty dollars an acre, and the gross yield was $250, leaving close to two hundred dollars in profit. By the standards of the 1850s, that was an excellent financial return. Morris stated that "if it were not for the sudden and tremendous fall in prices to which tomatoes are subject soon after they come into market, growers might become rich in a few years."[74] He preferred raising tomatoes at thirty-seven cents a bushel to potatoes at seventy-five cents because he did not have to dig into the ground for the tomatoes.

Tomatoes were picked green and transported great distances. From southern states, tomatoes were shipped northward weeks before they were available locally. By the late 1840s truck farms in Maryland, Virginia, and Georgia sent tomatoes to Philadelphia, New York, and Boston. This was such a lucrative trade that New Jersey farmers migrated to Virginia during the 1850s to grow tomatoes for export to northern states. Similarly, the business was so profitable before the Civil War that tomatoes were grown in Bermuda and exported to the middle and northern states in May and the early summer months. Although this competition seriously interfered with the profits of New Jersey tomato farmers, it did not destroy them. Edmund Morris reported that when prices fell during the summer, "the Southern growers could not afford the cost of delivery here, and thus left us in undisputed possession of the market. But, as a general rule, the Virginia competitors invariably obtained the highest prices. A great portion of their several crops, however, perished on their hands; because, as they had no market here when prices fell, so the scanty population around them afforded none at home."[75]

By 1851 Patrick Neill affirmed that tomatoes were a crop in immense cultivation in the middle states. Around Philadelphia the tomato was considered the "prince of the vegetable market" and was "an object of extreme field cultivation." Edmund Morris reported that a vast area was planted with tomatoes. By 1858 thousands of acres were cultivated to supply the demands of large cities.[76]

This demand skyrocketed during the Civil War, mainly due to the vast expansion of the canning industry. After the war New Jersey, Delaware, Maryland, Pennsylvania, Indiana, and Ohio became major tomato growing and canning centers. Due to the high demand for tomatoes, seed became scarce in New Jersey after the war. Seeds for early tomatoes sold for six dollars per pint.[77] New Jersey farmers began experimenting to develop even earlier varieties of tomatoes. For instance, C. T. Crolic of Plainfield, New Jersey, crossed the French tree tomato with the Early Round variety and produced a fruit that ripened by June 22.[78] Still others bred particularly large tomato varieties. An Irvington, New Jersey, gardener raised one with a circumference of almost twenty-six inches. It weighed three pounds.[79]

These experiments and boasts were critical. Despite the expansion of tomato consumption prior to the Civil War, the tomato had botanically changed very little since its initial introduction. The original tomato fruit was two-celled. Pre-Columbian farmers had developed a tomato with a large, lumpy, multicelled fruit. Mesoamericans selected and nurtured other variations. By the time mainstream Americans cultivated tomatoes, the fruit came in a wide variety of shapes, sizes, and colors. Despite this extensive diversity in tomatoes, the *American Agriculturist* maintained in 1848 that few varieties were esteemed in the United States.[80] William Chorlton described tomatoes that large growers sold as "thick skinned, hollow subjects, which are too often seen on the huckster's stall, which 'bounce' like a foot-ball."[81]

NEW TOMATOES

New varieties were imported into the United States from France, the United Kingdom, Italy, and other countries. This dramatically increased and improved the gene pool. One imported tomato was a variety purportedly brought back from the Pacific in 1841. The *U.S. Gazette* reported that the American Exploring Expedition in the South Pacific had run across a variety they called the Fejee tomato. A sailor sent Fejee tomato seeds back to a friend in Philadelphia, while Charles Wilkes, the captain of the expedition, dispatched them to Secretary of the Navy James Pauling. He evidently dispersed them. These particular Fejee seeds had no discernable effect upon tomato culture and apparently disappeared after a few years. The name Fejee was later applied to another variety, probably imported from Italy, which became popular during the Civil War. Wherever it came from, its fame was unequaled in the United States. The fruit was a large, solid, rough, very heavy, productive, purple-skinned sort. Seedsman Burnet Landreth of Pennsylvania believed that the Fejee was the ancestor of all other purple sorts. Now, "purple" used in reference to tomatoes really means either a deep red or a pink color.[82]

Farmers, gardeners, and seedsmen used newly introduced varieties and slowly bred other tomatoes in the quest to develop round, smooth-skinned fruit with solid flesh. The benefits of these breeding efforts first became apparent just before the Civil War. Fearing Burr's *Field and Garden Vegetables of America,* initially published in 1863, reflected his experience as a seedsman and gardener in Massachusetts during the late 1850s. He listed twenty-two tomato varieties, only one of which is not now considered within the genus *Lycopersicon*. The 1865 edition of his book listed two additional varieties.[83]

In 1858 Henry Tilden of Davenport, Iowa, discovered a tomato growing in a field: it was solid and prolific, but dwarf. He christened it the Tilden tomato. In 1864, Henry Tilden claimed to have grossed six hundred dollars from the sale of fruit from only one acre planted with his new tomato variety. It was better than other varieties on the market at the time and was released to the public by seedsman Apollos W. Harrison of Philadelphia in 1865. Of the Tilden tomato, a writer in the *American Gardener* wrote: "It was red, inclined to grow long fruited, smooth and without seams or creases, large, but not so early as Cook's Favorite, but much more strong."[84]

Cook's Favorite originated in Burlington County, New Jersey. It was introduced by seedsmen in 1864. Its plants were strong and vigorous, with fine, broad, light green foliage. It yielded abundant crops of smooth, handsome fruit, which were of deep red color, medium size, and roundish or oval in shape. Its smooth flesh was firm and remarkably solid, containing little water and few seeds as compared with other varieties. The fruit kept longer after being gathered and rarely had a cavity or hard, unripe parts at the center. Cook's Favorite was popular in the middle states, where it was extensively grown in the vicinity of New York, southern New Jersey, and Philadelphia by market gardeners and for supplying the large canning establishments. Ten years later, it was rechristened the Canada Victor.[85]

The first successful commercial tomato variety was developed by Dr. T. J. Hand, originally from the town of Sing Sing, New York. During the 1850s he began crossing the small cherry tomato with several larger lumpy varieties. His efforts were rewarded when he produced a tomato with a solid mass of flesh and juice, small seeds, and smooth skin. Under the name Trophy tomato, its success was unrivaled after the Civil War. Its promoter, Colonel George Waring, sold seeds for the unheard-of price of twenty-five cents apiece.[86] The Trophy was an improvement upon other varieties at the time, but far more important was Waring's demonstration that it was possible to make a fortune selling tomato varieties. This encouraged others to redouble their efforts to breed new ones.

Alexander Livingston of Columbus, Ohio, began experimenting with tomato breeding in 1848 without much success. In 1865 he found a tomato plant with heavy foliage that produced a prolific, uniformly round fruit. After five years of work, Livingston was satisfied with his product and marketed it under the name of Livingston's Paragon. While the Paragon was a good variety, Livingston continued searching for new varieties. He found another plant, whose fruit ripened early with solid flesh. Livingston developed this plant just as he had developed the Paragon. In 1875 he released it under the name of Livingston's Acme; it did not have a green core and possessed few seeds.[87]

Not just any tomato could be canned. Out of a field of Paragon tomatoes, Livingston located a specimen that he believed approximated what canners needed in a tomato. After improving it, he introduced the tomato in 1883 as Livingston's Favorite. It was an early, blood-red, smooth, solid, meaty, large, and prolific tomato that ripened evenly.[88] Livingston was always on the lookout for new varieties with unique characteristics. In 1885 he obtained a specimen from a market gardener near Columbus that appeared particularly promising, for it produced a thick, solid, red fruit. It was shaped like the Favorite. Livingston continued his experiments and released it in 1889. As the fruit weighed more than any other of his varieties, he called it the New Stone.

Alexander Livingston wrote down his tomato experiences and published them as *Livingston and the Tomato* (Columbus, 1894). The book was the first comprehensive book about tomatoes published in America. It chronicled the rapid conversion of the tomato from a ribbed, hard-cored, frequently hollow object into many new varieties that provided the base for the tomato's development during the twentieth century. The book also included extensive information about Livingston's methods and his results, about how to cultivate tomatoes, and about how to respond to tomato diseases and pests. Finally, Livingston's book featured over sixty tomato recipes, including four for canning tomatoes and five for making tomato soup. Alexander died in November 1898, but the company prospered under the control of his sons and grandsons, who continued to develop new tomato varieties. Twentieth-century Livingston Seed Company varieties included the Globe in 1906, which was a cross between Livingston's New Stone and the Ponderosa. In 1917 a horticulturalist in the U.S. Department of Agriculture crossed the Globe with the Marvel—a French variety—and produced the Marglobe.[89]

Several new breeds originated or were developed in New Jersey. The King of Earlies, for instance, originated with Theodore F. Baker of Cumberland, New Jersey. It was early, red, and medium size.[90] John C. Gardener, former gardener for Pierre Lorillard of Jobstown, New Jersey, crossed Livingston's Acme with the Perfection; the result, the Lorillard, was introduced by A. D. Cowan & Company, New York, about 1889. It was of medium size, depressed globular in shape, and vermillion or scarlet in color, changing to a bluish tint when fully mature. It had four or five remarkably uniform fruits to a bunch, ripened evenly, and was very firm and heavy. It was suitable for shipping.[91] The Thorburn Seed Company introduced the smooth, solid-fleshed New Jersey tomato variety in 1889.

Peter Henderson of South Bergen, New Jersey, was a leading seedsman who experimented extensively with tomatoes. During the late 1860s and early 1870s, he bred and marketed several tomato varieties.[92] According to Henderson, the tomato plant required a peculiar soil and location to produce early. He often saw "a difference of two weeks in the ripening of this fruit from the same sized plants, planted the same day, in situations only half a mile apart, but on entirely different soils, those of the light sandy soil selling by their earliness at 44 per bushel, those on the stiff, clayey soil, two weeks later a drug at one-forth that price."

Henderson believed that the tomato was only "profitable in warm southerly portions of the country, where there are rapid facilities to get them to the Northern market." For instance, Henderson pointed out that the tomato crop raised around Baltimore was supplanted by the tomatoes raised in the vicinity of Norfolk, "the Baltimore crop again, in turn, supplanting that of New York." He stopped growing tomatoes because he was "convinced that it was far from profitable in this section, although there is no doubt that in warmer latitudes, within transporting distance (say sixty hours) of our large cities, it must be highly so."[93]

Shortly after Henderson's dire pronouncements, tomatoes became one of New Jersey's predominant crops. Some found profit in growing tomatoes under glass. J. S. Lippincott of Haddonfield, New Jersey, described the raising of tomatoes in hotbeds in Camden by "Pea Shore" truck farmers who commenced work about February 20. They constructed hot beds composed of "fourteen inches of good, fresh horse manure, well shaken up and then slightly compressed." On the horse manure was placed about four inches of soil, and early tomato seeds were sown in hills. Some successful raisers of early tomatoes carefully selected "the earliest, smoothest, largest and fairest for seed, year after year, and thus secure[d] a variety which can compete with any to be had at the seed stores." Early tomato seeds were generally scarce and at times commanded as much as six dollars a pint. The seeds were "sown rather thickly in the hotbeds and the plants are carefully watched, aired on proper occasions at mid-day, and covered with old hay during stormy or cold weather." When the plants attained the height of four or five inches, they were "carefully drawn and transferred to a cold frame, also covered with glass, and having a few inches of rich, old soil and old manure beneath." Under glass again, the plants were "carefully watched, ventilated, covered with hay in windy, stormy, or cold spells." As soon as all danger from frost was past, generally about the first week of May, the plants were "ready, after a slight exposure without glass by day to endure the trials of the outer world."[94] While New Jersey farmers continued to make profits selling fresh tomatoes to large urban areas, the most phenomenal growth was caused by the rise of the tomato canning industry.

■ CHAPTER 3 ■

The Well-Preserved Tomato

From our prehistoric beginnings, humans have tackled the problem of how to preserve surplus foods for use in times of want. Hunter-gatherers saved food for short durations, but it was difficult to cart around preserved foods while moving from one location to another. The challenges of preserving food for extended periods of time became acute when the Neolithic revolution tied farmers down to particular locations. In regions where year-round growing of crops was impossible, food had to be preserved from one harvest to the next—usually almost a year. Numerous ways were developed, from drying food in the sun to salting food in containers. Which method was used depended in part on climatic conditions. In hot and dry regions, such as Egypt, sun-dried foods predominated. In wet and humid regions, such as in Southeast Asia, liquid forms of brining evolved.

The availability and quality of salt was another factor. It is no surprise that salt mining and extraction was one of the earliest industries, and salt was a major item of trade thousands of years ago.[1] The use of salt for preserving was common throughout the ancient world. Salt retards the growth of unwanted molds. At certain levels, salt creates an environment that encourages fermentation, which transforms tissue into simpler compounds through the action of enzymes produced by microorganisms; salt specifically encourages microorganisms that produce lactic acid. These chemical changes affect the tissue's texture and create characteristic flavors and aromas. Fermented products are often more digestible, nutritious, and flavorful.[2] Brining or pickling is a method using salt water to immerse the food to be preserved. Additives are mixed with the brine to create particular flavoring; in Europe, for example, vinegar and aromatics are added to the pickle.

In addition to salting and fermenting, other preserving techniques were employed in Europe, including smoking, sugaring, and freezing by packing food in ice or snow.[3] These techniques were used for hundreds of years with little change.

APPERTIZING

In the late seventeenth century, cookery books and medical works mentioned a new means of preservation: potting fruit in earthenware containers sealed with butter, lard, gum, meal, or charcoal.[4] William Salmon's *Family Directory* (London, 1705) featured a recipe for preserving fruit in jars. E. Smith's *Compleat Housewife* (London, 1727) corked bottles to keep out air. Smith's recipe was reprinted by several cookery books in the eighteenth century, including Hannah Glasse's *Art of Cooking* (London, 1747). Early techniques did not completely prevent air from coming in contact with the preserved food and thus resulted in substantial spoilage, but they were promising enough to gain the attention of several scientists.

At about the same time, scientists from several countries explored bottling techniques. British naturalist John Needham concluded in 1745 that "spontaneous generation" of life was possible. To prove it, he boiled and bottled meat. Several weeks later, he found microorganisms growing in the bottle. Twenty years later, the Italian naturalist Lazzaro Spallanzani demonstrated the fallacy of Needham's experiment, which had not sterilized the bottle. Spallanzani's experiments demonstrated that when the bottle was sterilized and air excluded, spontaneous generation did not occur. The Swedish chemist Karl Wilhelm Scheele independently sterilized vinegar by boiling it and sealing it in hot bottles from which the air had been exhausted. But these demonstrations that food could be preserved by boiling and hermetically sealing to exclude air were experiments, and practitioners did not follow in their footsteps until Nicholas Appert did so in the late eighteenth century. Appert was born in 1749 in Chalons-sur-Marne, France. He served as a superintendent of distilleries and breweries, a confectioner, and a provisioner to the ducal house of Christian IV.[5]

Shortly after the French Revolution, France became embroiled in a series of wars with various other European countries, particularly Great Britain. The British frequently established blockades of French-controlled ports. These blockades stopped the importation of sugar from islands of Martinique and Guadaloupe in the Caribbean. Sugar was used for many purposes, but particularly by confectioners and those engaged in preserving foods. In 1795 the French government offered a prize for an improved method of conserving food, and Nicholas Appert began to experiment. He first examined traditional techniques, which he divided into two major categories: *dessication*—drying or smoking; and *mingling*—sugaring, pickling, and salting. Appert concluded that these methods were beset with problems. Dessication destroyed the flavor and texture of the food. Mingling required the use of sugar or vinegar, which were expensive, and salt, which was believed to cause scurvy at the time. Mingling methods also greatly altered the taste of the food.[6]

Appert pursued the methods explored earlier by British, Swedish, and Italian scientists. In particular he may well have been aware of the experiments of Spallanzani, who preserved food by boiling it and preventing its contact with air. Appert accomplished this by packing fruits, vegetables, meats, and other foods into wide-mouthed glass bottles, sealed with a stopper made of cork tightly wired to the bottle, and heating them for various lengths of time in a bath of boiling water. Afterwards he inspected the seal to make sure that it was not broken. In 1804 the French Council of Health reported that Appert's food was in excellent condition after bottling. Appert presented several bottles of broth to the naval prefect at Brest. The prefect reported that after three months, the bottles of broth were still good.[7] In 1805 Appert constructed a small bottling factory at Massy, near Paris, where his food was grown. This permitted bottling shortly after the fruit or vegetables were picked, thus preventing delay and deterioration in quality of the food. Grimod de la Reynière, a French gastronome and author, praised Appert's bottled food and was happy to eat spring vegetables "in the heart of winter."[8]

Appert's experiments became known in Great Britain, where his methods were widely adopted. Based on Appert's work, for instance, Thomas Saddington authored a paper titled "A Method of Preserving Fruits Without Sugar for Home or Sea Stores" in 1807. Like Appert, Saddington heated fruit in loosely corked bottles in a water bath for an hour. The bottles were then recorked and sealed with cement. Saddington's article was reprinted in the United States through the 1830s.[9]

In 1808 Sir Humphry Davy, an English chemist, discovered that by adding calcium chloride to boiling water the temperature could be increased to 240°F. At the time this discovery was not particularly useful, as the glass used in canning was not strong enough to survive prolonged exposure to high temperatures. Davy's discovery was promptly forgotten.[10]

Meanwhile, Appert continued his experiments in France. On June 21, 1809, the Society for the Encouragement of National Industry requested solutions for means to preserve food without sugar, which continued to be scarce due to the British blockade. Appert submitted an account of his experiments, and on January 30, 1810, the minister of the interior requested a report based on his fifteen years of experiments.[11] Appert was awarded twelve thousand francs for his efforts. He used the money to rapidly expand his factory. In 1814 it was destroyed by the allies during the Napoleonic Wars. He was forced to sell land to repay his mortgage, but he continued to bottle goods. He reestablished his business after the war and continued his experiments until his death in 1841.[12]

His work was published in 1810 under the title *L'Art de conserver pendant plusiers années toutes les substances animals et végétales.* In October of the same year, it was translated into German; a year later, into English. The British translation of Appert's work, *The Art of Preserving,* was reprinted in the United States in 1812.[13] Appert's book described his experiments and offered specific formulas for the preservation of boiled meats, condensed milk, fruits, and vegetables, including tomatoes. His tomato recipe was as follows:

> The tomatoes are gathered well matured, when they have acquired their fine color. After they have been washed and drained, they are cut into pieces and put to soften in a well tined copper vessel. When they have been softened and reduced a third of their volume, they were strained, the decoction was replaced on the stove, and concentrated so that there remained only a third of its total volume; after cooling in stoneware dishes, it was put directly into bottles, etc., so as to give a good boiling only in the water-bath etc.[14]

Appert made no statement regarding the culinary usage of tomatoes after uncorking, other than to say he employed them in the same way as he did fresh tomatoes in season. While tomatoes probably had been used previously in Provence and southern France, they were not commonly used in Paris or northern France until the nineteenth century.[15] Tomato recipes did not regularly appear in French cookbooks until the 1830s, and even then tomatoes were mainly employed as ingredients in sauces and soups.

Appert's early experiments mainly preserved foods in glass bottles. He later tried tin but was generally unsuccessful, mainly due to the state of tin work in France. At the time Wales was the main center of the tin trade. The first use of metal cans to preserve food is credited to the Royal Laboratory in Greenwich, which made canisters as early as the mid-eighteenth century. At about the same time, salted and kippered fish were canned in Holland. This form of canning used a different preservation technique than that presented by Appert; the fish had not been sterilized but was preserved with brine and smoke.

Shortly after Appert's work was published in 1810, Augustus de Heine patented the use of iron cans, and Peter Durand the use of tin ones, for preserving foods. Neither Heine nor Durand, however, launched a commercial canning operation. Brian Donkin bought Durand's patent and went into the business of canning foods with John Hall, founder of the Dartford Iron Works. Three years later Hall and Donkin were selling canned meat. In 1814 their goods, which by then included vegetable soup, were taken on H.M.S. *Isabella* and *Alexander* for the voyage to Baffin Bay. The sailors received a pound of vegetable soup per week.

The Admiralty became enamored of canned goods. By 1818 British firms tinned many foods, including 5,498 one-pound cans of bouillon and 4,484 one-pound cans of vegetable soup. These soups generally elicited high praise. In 1820 an assistant surgeon of the H.M.S. *Griper* reported: "The soups I consider particularly excellent, especially as I have every reason to believe that the antiscorbutic quality of the vegetable is not injured in its preparation."[16] At a time when scurvy was still a problem on board all ships traveling long distances, this was genuine praise.

Initially, tin canisters were constructed by hand. After the food item was placed in the can, the lid was soldered on. In 1823 the vent hole was invented, making it possible for the food product to be inserted through the lid. The can was then placed in a water bath and sterilized. A circular piece of tin was soldered over the vent hole.[17] Early canned foods were extremely expensive and were sold as gourmet food or used for long sea voyages, such as the exploration of the Arctic. The contents of some cans left from these early expeditions were analyzed 114 years later. There was no rancidity, and surprisingly, the vitamin D survived.[18]

AMERICAN CANS

In the United States, food preservation started at a much slower pace. It was largely based on the migration of English canners, such as William Underwood, who had served a pickling and preserving apprenticeship with London's Mackey & Company. He migrated to New Orleans in 1817, but he was not satisfied. Out of money, he started off across the country on foot, arriving two years later in Boston, where he created the firm later known as the William Underwood Company. By 1821 he was shipping plums, quinces, currants, barberries, cranberries, pickles, ketchup, sauces, jellies, and jams canned in glass to South America. By 1825 he was packing tomatoes in bottles. By 1839 he was preserving food using tin canisters.[19]

Evidence indicates that American canners, perhaps William Underwood, were the first to use the word *can*. Most likely it was initially an abbreviation for *canister*. The word *can*, meaning canister, was used in the United States by 1839. British canners did not adopt this Americanism and continued to use the words *canister, tin,* or *tinned foods*.[20]

Underwood may not have been the first to can with tin in America. In 1838 the *Horticultural Register* credited the first canning use of tin to "a French gentleman, some years since a superintendent to the late Doctor Hosack, of Hyde Park, N.Y., who practised it very successfully at that place." This unnamed Frenchman made canisters with "a piece of tin forming a tube of different dimensions, from eight inches to a foot in length. The width on the top is from three inches to six; on the bottom from four to eight."[21]

Ezra Dagget and his son-in-law, Thomas Kensett, began using these new tin cans in packing salmon, lobsters, and oysters in New York City about 1819. Subsequently Kensett moved to Baltimore, where he packed other animal and vegetable substances. The cans were made entirely by hand. The food was placed in the can, the lid soldered in place, and the can put into the water bath; a hole in the lid provided a means to exhaust the can. When the bath was finished, the hole was soldered and the can was totally sealed.[22] Essentially this system was employed for the next eighty years.

Tomatoes were still mainly packed in glass or earthenware jars; preserving them in tin remained experimental. Thomas B. Smith of Philadelphia reportedly tinned tomatoes in 1838.[23] Other parties claimed to have canned tomatoes during the early 1840s. George F. Lewis, for instance, packed raw tomatoes in cans after scalding them to remove the skins. Of these early efforts, none was particularly successful, and only small quantities of tomatoes were canned. Neither did these early tomato canners make any special effort to introduce their products to the general public or food retailers.

CROSBY'S TOMATOES

Harrison W. Crosby of Jamesburg, New Jersey, is credited as the first to popularize canned tomatoes. While serving as the steward of Lafayette College in Easton, Pennsylvania, Crosby filled six tin pails with tomatoes in 1847. He expanded his operation during the following year. According to Crosby, he processed tomatoes in an ordinary wash boiler; "afterwards I put three iron sinks, about 30 inches long by 18 inches wide, in brick work, which admitted my using 4 feet wood in firing. The tomatoes were peeled, put in cans cold and bathed thirty minutes, with the vent open and kept open by a wire. They were then removed and capped with a piece of tin, with clipped corners." Cosby boiled the pails again, venting the steam through a small hole in the cap. After boiling, he sealed the hole in the cap with a tin disk. During the following year, he sold his canned tomatoes through hotel proprietors and saloon keepers. He also distributed them in New York City's Washington Market, where a man named Broas sold them for fifty cents a can. To gain visibility for his products, Crosby sent six pails to Queen Victoria and President James K. Polk in the fall of 1848. He also sent samples to two U.S. senators, newspaper editors, restaurants, and hotels. The *New York Tribune* declared of his canned tomatoes that "whatever the secret of their preparation, we are bound to acknowledge that their preservation has not impaired their flavor. They taste as they would have tasted when plucked from the vines."[24]

Soon others were packing tomatoes in tin canisters. In September 1849 John Ferguson, Philip D. Watson, and Reuben E. Hills contracted with the firm of McGhee, Harshman and Miller in Ohio to furnish three thousand quart cans of tomatoes for $225. The cans were shipped to San Francisco, where the gold rush had created an immense demand for food. It cost $113 to transport the cans from New Orleans to San Francisco by schooner. But it cost $130 to ship two thousand cans from San Francisco to Sacramento, where the retailer paid two thousand dollars for them. The cost to the consumer was undoubtedly much higher. What caused the high price for shipping the cans to Sacramento was the paucity of labor, as most Californians were panning for gold. As a scarcity of food engulfed California, astronomical prices were charged for canned goods and other foodstuffs at the retail end.[25]

The first located description of canning tomatoes was published in 1849 in Maine and read:

> Pour boiling water on ripe tomatos and peel them. Cut them up and put them into two quart tin canisters, each having a hole two inches in diameter in the top. After they are filled, have a circular piece soldered over the opening, leaving an anvil-hole. Set the canisters into a kettle of boiling water for twenty minutes. Stop the anvil-hole with a spile of pine—finally when they are cool, cut off the spile even with the tin, and drop a drop of melted sealing-wax over it. On opening your canisters you will have tomatoes in perfection at any time of year. N.B. Sprinkle a little salt and pepper on the fruit previous to packing in canisters.[26]

The *Southern Cultivator* offered a similar process in 1853. In addition the author pointed out that cans holding about half a gallon or less cost only twelve and a half cents each. The secret of canning successfully was to exclude all air and to store in a cool place. According to the author, when a dish of tomatoes that had been canned the previous year was compared with tomatoes picked recently, "no one could distinguish the one from the other." The author also believed that the general consumption of tomatoes "would promote the public health and morals."[27]

CIVIL WAR TOMATOES

These changes and their wide dissemination greatly enhanced the ability of the canning industry to respond to the needs generated by the American Civil War. Beginning in 1863, the federal government pumped money into the industry, and canneries sprang up across the northern states. This Civil War expansion meant that many more Americans had experience with canned goods. They were introduced to soldiers in the garrison or in the field, sailors in the fleets blockading southern cities or in Union gunboats along the rivers, and the wounded and sick in hospitals. This primed the pump for an upward spiral of demand, causing an increase in supply. Through former members of the army and navy, the general public became acquainted with canned goods, and demand for them exploded after the war was over.[28] Continued demand induced dramatic expansion of the canning industry. Technical advances brought lower prices, and canned foods became affordable to middle-class Americans. As the variety of canned goods grew and the price continued to decline, demand increased.

After the war, canned foods were commonly available in most markets. For instance, a grocer in Flemington, New Jersey, offered canned tomatoes as well as canned peaches, pineapples, blackberries, whortleberries, green corn, beans, peas, asparagus, spiced oysters, clams, lobsters, sardines, and salmon. Volume also rose rapidly. Only five million cans were put up annually in 1860; the figure was six times greater a decade later.[29]

This expansion was particularly reflected in the canning of tomatoes. In 1868 Edmund Morris reexamined the tomato fields of New Jersey. He reported that South Jersey was cross-hatched with tomato fields. Previously growers had made their money on the sale of early tomatoes, which sold for "four or five dollars per basket of three pecks." Farmers planted acres of tomatoes just to yield some fruit that matured early. As the main crop ripened, a glut on the market caused much to be thrown away.

> This large waste gave rise to a new and extensive business—that of canning in large manufactories. When the supply now becomes abundant, so that the tomatoes may be sold at low rates, these manufactories take the surplus. A single establishment in Philadelphia has employed four hundred women at a time for canning tomatoes and small fruits, and consumed in a single season $30,000 worth of sugar. There are three manufactories at Burlington, New Jersey, which employ six hundred persons, and disburse weekly several thousand dollars for wages. They consume the product of nearly a thousand acres, lying within three miles of Burlington.[30]

HOME CANNING

Home gardeners faced the same dilemma as did large tomato growers. At the end of the tomato season, gardeners were deluged with fruit. Canning with tin was difficult. The need for easier means of preserving food in the home led to the creation of a variety of new techniques and devices. In 1855 Robert Arthur took out a patent for a grooved-ring can. He created Arthur, Burnham & Company in Philadelphia and began producing "patent, air-tight, self-sealing cans and jars," which became known as "Arthur's cans." They had an open top with a lid that extended over the top part of the jar and was sealed with cement prepacked in its lip. According to *Godey's Lady's Book,* these were used by thousands of families, hotels, and boardinghouses.[31]

From the consumer's standpoint, the disadvantage of Arthur's cans was that they were not easily reusable. After the contents were removed, it was difficult to reglue the jars. Although Arthur did claim that other substances could be packed in the groove, such as cloth, leather, or even newspapers, one wonders how effective the seal was. His jars were soon overshadowed by the invention of New Jersey–born John L. Mason, who had set up a metalworking shop on Canal Street in New York. On November 30, 1858, when he was twenty-six years old, Mason patented the self-sealing zinc lid and glass jar. The screw-on lid greatly simplified the canning process and made the jars genuinely reusable. It revolutionized fruit and vegetable preservation in the home. As the jars were easy to use and comparatively inexpensive to produce, their popularity soared. By 1860 Mason jars were shipped throughout the United States.[32]

After the Civil War, home canning rapidly expanded, but not everyone was satisfied with the Mason jar. From Cecil County, Maryland, came the report that tomatoes were difficult to put up in glass. A correspondent to the *Country Gentleman,* identified only as T.R., reported that "some use stone jars, crocks, &c., and think they succeed pretty well," but he frequently found traces of fermentation in tomatoes. T.R. preferred using tin to put up tomatoes. In 1865 he "filled one dozen glass jars from a kettle of the hot tomatoes, and scalded them carefully. The remainder were put in tin cans and soldered up." Those packed in glass spoiled, while those canned in tin "kept in good order with only the usual percentage of loss." Tin cans preserved the contents "just as good for cooking as the day they were taken from the vines."[33]

Despite this laborious procedure required in home canning, T.R. believed that it was as cost-effective as bottling, since much of the glass broke. Also, the tin cans could be reused; they would "last for years if they are always carefully opened with the hot iron, and immediately after being emptied are rinsed and dried. I know a housekeeper who has had them in use for five or six years, and they are still in good order and ready for further service."[34]

THE JERSEY CANNER

Almost from the beginning, factories in New Jersey canned tomatoes, including Aldrich & Yerkes of Moorestown and Lorin J. Wicks and J. W.

Stout of Bridgeton. In 1848 and 1849 Bamford, Bellville & Lewis of Trenton packed twelve hundred cans. These tomatoes were packed hot, after being cooked in open copper kettles on a large range. Both lots spoiled, and the partnership broke up. Bamford persevered and in 1850 packed fifteen thousand cans, increasing the quantity each year until the total pack reached a hundred thousand cans by 1855. Harry Evans established a factory for Kensett and Company in Newark, New Jersey, where tomatoes were packed for Elisha Kane's Arctic Expedition in 1851.[35]

John Bucklin established a factory at the North American Phalanx in Red Bank. Bucklin bottled tomatoes during the late 1840s. In 1851 he experimented with canning tomatoes. To examine longevity, he saved one case. Confident in the process, Bucklin began commercially canning tomatoes by 1855. With his son, Bucklin later founded the firm of J. & W. S. Bucklin, which enjoyed an excellent reputation in the industry, packing tomatoes for nationally known firms such as E. C. Hazard & Company of Shrewsbury, New Jersey. More significant was Bucklin's invention of the tomato filler, a machine to fill cans automatically, which revolutionized the tomato canning industry during the 1880s. The case of tomatoes that Bucklin canned in 1851 was opened in 1891. Bucklin presided at this tomato party in person. The tomatoes were in perfect condition and were "hugely enjoyed by all present, particularly by Mr. John Bucklin. The wood case or box the tins were packed in had crumbled away through dry rot, but the tins and their contents were in perfect condition."[36]

In New Jersey, suppliers of fresh tomatoes had multiplied rapidly before the Civil War. Growers planted thousands of acres and made their main profit on the tomatoes during the first few weeks of the season. As tomatoes matured, the price dropped so low that it was cheaper to plow them under rather than pick them. When the price declined to its lowest point, canners started purchasing tomatoes.

Prior to 1860 factories were limited to manufacturing two or three thousand cans at most per day, mainly because some cans had to be heated for five or six hours to prevent spoilage. In 1860 this changed. Isaac Solomon, the manager of a tomato canning factory in Baltimore, rediscovered Sir Humphrey Davy's experiments with calcium chloride, which, when added to the water, permitted the temperature to increase to 240°F. Higher temperatures meant that sterilization occurred much faster. Processing time was reduced to between twenty-five and forty minutes. Reduced time meant less expenditure on heating and a vast increase in production. Using these new methods, canners produced as many as twenty thousand cans per day.[37] Lower expenses and increased supply meant a steady decrease in the price of canned goods.

When the federal government let contracts throughout the northern states for food to feed the military in 1863, canneries sprang up in truck-gardening areas. Tomato growing had been extremely successful east of the Delaware River across from Philadelphia. It was an ideal location for tomato canners, and in 1863 John M. Butterfuss of Lambertville, New Jersey, started packing "Hermetically Sealed Goods."[38] Edmund Morris visited Butterfuss's cannery and reported that it employed thirty people (mainly women) and produced fifty thousand tomato cans in a single season. The canner purchased tomatoes when the price dropped to twenty-five cents per bushel. At the high season, a hundred and fifty bushels arrived at the cannery every day, which required everyone to work far into the evening. Morris described the operation:

> The building in which the business is carried on was constructed expressly for it. At one end of the room in which the canning is done is a range of brick-work supporting three large boilers; and adjoining is another large boiler, in which the scalding is done. The tomatoes are first thrown into this scalder, and after remaining there a sufficient time, are thrown upon a long table, on each side of which are ten or twelve young women, who rapidly divest them of their leathery hides. The peeled tomatoes are then thrown into the boilers, where they remain until they are raised to a boiling heat, when they are rapidly poured into the cans, and these are carried to the tinmen, who, with a dexterity truly marvelous, place the caps upon them, and solder them down, when they are piled up to cool, after which they are labeled, and are ready for market. The rapidity and the system with which all this is done is most remarkable, one of the tinmen soldering nearly a hundred cans in an hour.[39]

Farther south in New Jersey, tomato canning got off to a dismal start. Tomato canneries were first launched during the 1850s. The best evidence sets the date of 1853 for the first factory in Salem. During the Civil War, more factories sprung up. They were in constant flux. Another canning factory in Salem was built in 1864 by Theophilus Patterson, Richard B. Ware, and Charles W. Casper. They produced fifty thousand cans of tomatoes, but the business was not profitable: tin plate was expensive, women were opposed to working in the factory, and few farmers were willing to grow tomatoes only to sell them for twenty-five cents per basket or twelve dollars per ton. The partnership dissolved at the end of 1864, but the business was continued by James K. Patterson and Ephraim J. Lloyd, and afterward by James K. Patterson and Owen L. Jones. In 1872 Patterson dropped out, Ayers came in, and the name changed to Jones and Ayers. Their annual output was about four million cans.[40] By 1875 New Jersey was rated as the state that produced the most canned tomatoes.[41]

Tomato canners received good visibility at the Philadelphia Centennial Exposition, which opened on May 19, 1876. Fairgoers flocked to the

236-acre site with its 180 buildings filled with exhibits—fifty of which were sponsored by other nations. When it closed six months later, over ten million visitors had passed through the gates. At the Exposition, canned tomatoes received awards. A canning company in Camden, New Jersey, named Anderson & Campbell received an award for excellence in preserving vegetables; its major product was canned Beefsteak tomatoes. Others displaying their tomato products were A. B. Roe of Greensborough, Maryland, who sold "Centennial Tomatoes," and the South Jersey Packing Company of Cedarville, which exhibited canned tomatoes.[42]

Although the 1876 Exposition proved a success for canned tomatoes, the year was a disaster for the tomato canning clan. The supply of canned tomatoes had far outstripped the demand for them. The problem had begun in 1874, when over one million cases of tomatoes containing twenty-four cans each were produced. Only 350,000 cases were produced the following year, but a national depression had set in. Consumers were not buying canned foods, which many considered luxury items. At the beginning of 1876, four hundred thousand cases manufactured in 1874 remained unsold. Hence, three quarters of a million cases of tomatoes were on the market. The depression deepened through 1876, reducing demand for canned goods still further. Inclement weather exacerbated the disaster in the fall. George Moore, a tomato grower in Cumberland County, recorded a short crop of less than eight tons, about one third to one half his usual harvest. As he was under contract, he received only forty-five cents per hundredweight. Others not under contract drove up the price of raw tomatoes. The price of previously canned tomatoes dropped below what manufacturers could afford to charge to produce new ones. In consequence, many tomato canneries folded.[43]

One effect of the tomato failure in 1876 was greater diversification of tomato growing areas outside of New Jersey. In particular, Maryland boosted its tomato production. By 1878 Maryland canned more tomatoes than New Jersey.[44] But this was reversed in 1880 when New Jersey produced over half a million cases, about a third of all the tomatoes canned in the United States.[45] For years Maryland and New Jersey seesawed back and forth as to which state canned the most. During the mid-1880s, New Jersey was firmly in the lead. In 1886 New Jersey canneries produced 710,133 cases. At this time this was almost one fourth of the entire total tomato pack in the United States, which was 2,815,048 cases.[46]

Although tomatoes were grown throughout the state, the canning industry increasingly centered in South Jersey. For instance, a single canning factory, John M. Butterfuss, produced 340,000 cans in 1884. By 1891 Butterfuss's plant was equipped with improved machinery and a sixty-horsepower steam engine and boiler. During the canning season he employed a force of two hundred workmen. His plant produced half a million cans annually.[47]

During this time tomatoes were the center of attention in most South Jersey communities. Farmers there greeted each other, "How are your tomatoes?" Their main concern was how the weather would affect the crop. Rivalries ran high, and, as one observer reported, "every scheme feasible and otherwise is adopted to get a great yield. But the vegetable is very coy and uncertain, and the most strenuous efforts to woo it to produce a large crop are often futile. Just as the farmer thinks he has the system of cultivation that will make him a twenty-ton-to-the-acre crop, that is about the time he gets five or six."[48] In general farmers sowed their seed in beds beginning about April 10 in the warmest and richest location. While these seeds were germinating, farmers manured the fields with compost and some phosphate. The plants were transplanted from the beds to the fields during the last week in May. Tomato picking began about August 10 and continued until frost hit about October 1.[49]

Most farms in southern New Jersey from Trenton to Cape May cultivated tomatoes. In the spring the region was filled with yellow tomato blossoms. In the fall fields were awash in the green of foliage and the red of ripening tomatoes. Wagons and carriages of every description filled the roads on their way to the canneries. The roads were virtually painted red with squashed tomatoes that fell from the wagons. Most towns had one or more canneries. During September and October, these communities were dominated by tomato canneries.[50]

The most profitable crops of tomatoes in New Jersey were grown in Salem County. In most of New Jersey, an acre of good land yielded about nine tons of tomatoes. At first many Salem farmers produced fifteen to twenty tons per acre, but over time this decreased. Originally farmers received twelve dollars per ton, but by the end of the 1880s they were paid at six dollars per ton. Salem farmers produced phenomenal crops in 1886 and 1890. Small factories packed two hundred thousand cans of tomatoes; larger ones, a million. One canning establishment in Salem manufactured almost two million cans, generating $150,000 in income. Several large canneries in Salem packed five million cans annually.[51]

The business was carried on by a peculiar system. The canners received, packed, and sold the tomatoes but paid growers when convenient, usually at the end of the year. As many packers were fly-by-night operations, some failed to pay the growers at all. There was no chance of redress. Still, tomato growing, "with its pleasures and drawbacks," as one participant wrote in 1891, remained popular: "The possibility and hope of getting a phenomenal yield spurs them on, and new schemes are tried year by year."[52]

During the decade of the 1880s, tomato production quadrupled. In New Jersey, there were seventy-three tomato canneries by 1890. Over fifteen thousand acres were devoted solely to raising tomatoes for canneries. The average yield per acre was about nine tons, with a minimum selling price of $6.50 per ton, delivered at the canneries. At least two thousand additional acres were dedicated to growing fresh tomatoes, which sold for twenty-five cents per thirty-pound basket. The total annual value of the crop to the farmers of New Jersey was over one million dollars.[53]

The best was yet to come. In 1892 tomato canning scored another phenomenal year. Factories, large and small, canned twelve million quart cans, generating $268,800 for the growers and $425,000 for the canners. Had it all been packed in railroad cars and shipped at the same time, the train would have stretched for eight miles.[54] By the 1890s the tomato was New Jersey's most profitable vegetable crop.

AFFLICTED TOMATOES

Despite this rapid growth, or perhaps because of it, not all boded well for the tomato industry in South Jersey. It faced two major challenges. The first concerned the labor force required to harvest and can this huge amount of tomatoes. From picking to canning, tomatoes required a vast army of seasonal workers, particularly during August and September. On large farms, the backbreaking work of picking tomatoes was accomplished by African-Americans supervised by Quakers. Some canneries only hired white women, others only black women. Larger factories had separate rooms for each group. Unlike the women, black and white men worked side by side in most factories.[55] Some factory positions required skilled labor, and there was little of this to go around. When the season ended, neither the pickers nor the cannery employees were needed.

The second challenge faced by tomato canners was sanitation. Some factories were more sanitary than others, but all had problems. For instance, workers who skinned the tomatoes might or might not observe proper sanitary procedures. Tomato peelings, cores, and other refuse often rotted on the cannery's floors. Perhaps because of the rising production of canned tomatoes, numerous problems began to be brought to the public's attention.[56] Reports of poisonous cans of tomatoes filled newspapers and magazines. For instance, the *New York Times* carried the account of a Dr. J. G. Johnson, who said that he was called to see a family in Brooklyn suffering because of a poisonous can of tomatoes. Immediately after eating it, the whole family "became very ill with burning pains at the pit of the stomach, dryness of the throat, and bright red eruptions of the skin, and one of them was in a state of coma." It was subsequently determined that the lid of the can had been soldered on with muriatic acid, and some had dropped into the can.[57] The can had been sold by Thurber, Whyland & Company, and its officials were charged with selling poisoned products. Although the judge concluded that Thurber, Whyland & Company had sold defective goods, the charge was dismissed because the plaintiff had discarded the can.[58] In another case, tomato paste from Daretown, New Jersey, intended to make soup, swarmed with unhealthy bacteria.[59]

Unlike products packaged in glass, the contents of cans could not be seen by consumers. Some canners took advantage of this and filled their products with ingredients other than tomatoes. For instance, one manufacturer ran out of tomatoes and substituted apples. Another filled a tomato can with a "miscellaneous assortment of vegetables, including a few green tomatoes and red paint, the whole having the appearance of ripe tomatoes." When agents investigated, they concluded that "red paint was used to a great extent in coloring worthless and unripe tomatoes and that other coloring matter also was used to give a ripe appearance to worthless canned vegetables of various kinds."[60]

These abuses spurred on the movement to enact pure food laws. The first attempt to pass a national pure food law took place in 1876. Legislation continued to be introduced into every session of the U.S. Congress thereafter. It passed the House of Representatives on two occasions, but the Senate failed to vote on the bills. Opposition to federal legislation came from agricultural interests fearful of losing markets, food processors opposed to federal control, and political conservatives opposed to governmental interference in the marketplace and to infringement upon the powers belonging to the states. In addition, there was no general public support for the legislation. Supporters also worked at the state level to pass legislation. The agitation for state and national pure food legislation disturbed food manufacturers across America. While reputable manufacturers stood for honest labeling of all food products and guaranteed that their products were "wholesome and free from deleterious, objectionable or unhealthy ingredients," the opposition to pure food legislation included major tomato-product manufacturers, including Joseph Campbell Preserve Company, Curtice Brothers, P. J. Ritter Conserve Company, Williams Brothers, and E. Prichard Company.[61] Initially, pure food advocates demanded only that manufacturers accurately label their products. In the words of one such advocate: "Then if people want to eat dyes and poisons, they alone are responsible."[62] Labeling laws were passed in many states. Under these laws, commissioners could declare products misbranded if chemical analysis proved that the product included ingredients not listed on the label.

During the early twentieth century, newspapers and magazines published scathing attacks against abuses in the food industry. These attacks encouraged the lobbying efforts of the pure food movement. As it became increasingly clear that a national food law would pass Congress, pure food advocates set their sights higher, agitating to ban harmful ingredients. Many canners testified against the passage of the national food law.[63] But support for the national pure food bill increased dramatically due to the publication of Upton Sinclair's *The Jungle* and similar muckraking exposés published in *McClure's, Ladies' Home Journal, Collier's*, and others. When the bill that became the Pure Food and Drug Act was debated in Congress, adulterated goods were displayed in the lobby of the House of Representatives. This display incorporated several hundred food samples purchased at grocery stores throughout the nation. Attached to each was an analysis of the adulterants contained in the product.[64] With the general support of the American people, the bill finally passed Congress in June 1906. It provided for the end of adulterants in foods and required truthful labeling of foods sold across state lines.

POOLING TOMATOES

Despite these problems, tomato processing dramatically expanded during the decades surrounding 1900, when more tomatoes were canned than any other fruit or vegetable. In 1885 more than two million cases containing twenty-four cans each were produced nationally. Within three years, more than five and a half million cases were produced. During the first few years of the twentieth century, the tomato industry boomed. In 1902 over nine million cases were packed nationwide—almost double the number of previous years. The following year this record was eclipsed when over ten million cases were packed.[65]

But in 1904 the total number of tomato cans manufactured dipped to about eight and a half million, and the price escalated to seventy cents per case. At the end of the year, rumors of a "combination of capitalists to control the canned tomato market" began to circulate about the industry. According to these rumors, a syndicate had pooled a million dollars and intended to purchase large quantities of cans of tomatoes to control the market and drive up prices. In an era before most large trusts had been broken, several syndicates tried to corner the market on commodities, such as gold and silver. No one had yet attempted to control the price of canned goods. In 1905 the tomato crop was one of the poorest in years, and tomato canners turned out only five and a half million cans. The syndicate assumed the price would increase and commenced purchasing and hoarding the best quality of canned tomatoes. While almost anyone could pack tomatoes, quality was an important issue.

Only those canners with the modern equipment could pack the best tomatoes, and respectable grocers only stocked the highest quality to avoid handling returns by irate customers.[66]

Supported by bankers and brokers, the syndicate was composed of three principal individuals: Willard C. Rouse, Charles S. Crary, and Frank Van Camp. Rouse, a Maryland canner, was considered "one of the shrewdest traders in the industry." Crary was president of the Charles S. Crary Company of Streator, Illinois; at the time, he was also president of the National Canners Association. Frank Van Camp was a major canner based in Indianapolis. His company had been launched by his father, Gilbert C. Van Camp, who had started packing fruits and vegetables in the back of his Indianapolis Fruit House Grocery in 1861. Among his processed products were both beans and tomato sauce. In 1864 Van Camp combined the two, scoring an immediate hit with consumers—except in New England, where baked beans without tomatoes remained the favorite dish. The business thrived, and in 1882 it was incorporated as G. C. Van Camp & Son. Gilbert Van Camp died in 1900, and Frank took over the canning empire. Within the next few years, the company diversified and offered many other products, including canned soup.[67]

Despite a rise in canned tomato prices in 1905, the market was difficult to corner with just one million dollars. When the price went up to $1.15 per dozen cans—forty-five cents more than a year before—brokers and wholesale grocers refused to buy canned tomatoes. In part because the price for tomatoes was high, growers planted more tomatoes in the spring of 1906, and everyone decided to can them. The weather cooperated, and a bumper crop of tomatoes was forecast. In the fall, canners packed 8,631,138 cases of tomatoes. As the deluge of tomatoes poured into the market, the syndicate suffered but decided to hold on to its hoard and gamble that the following year's tomato crop would be smaller. But the following year yielded the largest tomato crop the nation had ever seen, and the industry canned almost thirteen million cases. The price for canned tomatoes declined precipitously, and the syndicate had to release its hoard. This depressed prices further. Rouse retired. Crary went out of the tomato canning business but later emerged as the first commercial canner of asparagus. The banks that held the notes foreclosed on Van Camp, and the company passed out of Frank's hands, becoming Van Camp's, Inc. Frank decamped for California, where he went into business with his son packing tuna.[68]

This marked the end of attempts to corner the canned tomato market. The Tomato Pool was defeated by nature's bountiful harvest and the ability of almost anyone to become a tomato canner. It was still a low-cost operation requiring a relatively modest investment in equipment, but there were signs that the times were changing. More efficient equipment being installed in larger canneries would enable them to pack millions of cans per day. The era of the small packer was drawing to a close. But the amount of tomato cans packed annually continued to increase. By 1917 the United States packed over fifteen million twenty-four-can cases. Canned tomatoes could be used in many ways: making soup was one of the most common.[69]

■ CHAPTER 4 ■

The Tomato in the Soup

The early culinary use of tomatoes in Britain and America was as one ingredient among many in soups. Tomatoes provided coloring, an acidic flavor, and a juicy texture unmatched by other fruits or vegetables. While references to tomatoes in soups appeared in English and American medical, agricultural, and botanical works, early recipes titled Tomato Soup were really vegetable soups. The Tomato Soup recipe in John Conrade Cooke's *Cookery and Confectionary* (London, 1824) contained tomatoes along with onions, carrots, celery, and beef broth seasoned with salt and cayenne pepper. In this case, the tomatoes' main function was to provide color. If the soup was not red enough, Cooke recommended adding beet juice.[1] Ester Copley's recipe in *The Housekeeper's Guide* (London, 1834) had similar ingredients, but the quantities had altered significantly. Her recipe increased the tomato content and decreased the quantity of other ingredients. In fact this was a move away from a vegetable soup and toward a purée dominated by the tomatoes.[2]

AMERICAN TOMATA SOUP

Most early American tomato recipes derived from Western European culinary traditions, but at least one popular recipe originated in Turkey. James Dekay visited Turkey and upon his return announced that sun-dried tomatoes could be concentrated in a thick paste. Dekay was surprised to note that "a bit not larger than a Lima bean will be sufficient to flavour the soup of a family of 20 persons; and a much smaller quantity for sauces. A small pot which I brought with me, containing about half a pint, lasted my family more than a year and we used it very freely."[3] This process was promptly simplified by pressing ripe tomato pulp through a hair sieve, spreading "it thin on plates or tins so that it may become quite dry, or it may be dried in a moderately heated oven."

According to Dekay, a small piece of this dried pulp flavored a tureen of soup. Presumably, other ingredients were added. But perhaps not, as the recipe was frequently reprinted in American newspapers and farming periodicals without any mention of additional components.[4]

The first American tomato soup recipe, titled Tomata Soup, was located in N.K.M. Lee's *Cook's Own Book*. It was likely borrowed from a British source. Similar to Cooke's earlier recipe, it consisted of three large carrots, three heads of celery, four large onions, two large turnips, butter, a pound of ham, three quarts of brown gravy, and "eight or ten ripe tomatas." The pulp was sieved out, and the soup was served with diced, fried bread.[5]

Lee's recipe was succeeded by others that further increased the amount of tomatoes and decreased other vegetables. Edward James Hooper's *Practical Farmer, Gardener and Housewife* included three recipes for tomato soup. One contained stock, water, eight or ten tomatoes, two sliced onions, a little okra, a spoonful of flour, and spices. Hooper announced that "epicures require a little cayenne."[6] Unlike previous tomato soup recipes, this one was not sieved, and hence resulted in a pulpy mixture. Hooper's recipe for Tomato Mutton Soup was similar, with the addition of a "*rack* of good wedder mutton." Hooper proclaimed that the tomato had "become a great favorite sliced and seasoned as we do the cucumber, and has the advantage of being quite wholesome." Americans, said Hooper ending with a flourish, were "stigmatised as the *worst cooks in the world!* It may be so; for in my peregrinations I have often sat down to a plentiful table with scarcely *an eatable* article on it; few of our cooks have *learnt any thing;* no, not even to *boil a potatoe* or bake a johneycake *properly!*"[7]

Despite such comments, soup flourished among American cooks. Tomato soup recipes proliferated during the 1840s. Lettice Bryan featured two similar recipes in her *Kentucky Housewife* (Cincinnati, 1841). Both were really purées consisting of two quarts of tomatoes dissolved in two quarts of broth. The mixture was strained and a few more tomatoes added, along with butter, pepper and salt. It was served with dry toast or crackers.[8] Other recipes were composed of tomatoes and fried onions in stock.[9] The first Jewish cookbook published in the English language, *The Jewish Manual* (London, 1846), also featured a similar recipe. Although it was published anonymously, its author has been identified as Judith Montefiore. Montefiore also included an intriguing recipe for Dry Tomato Soup, consisting of similar ingredients beaten to the consistency of a pudding. Eggs and bread were added, and the entire dish was baked until brown.[10]

Tomato soup recipes commonly appeared in cookbooks about the mid-nineteenth century. Subsequent recipes made but minor changes in the basic ingredients or methods.

FANCY TOMATO SOUP

Before the Civil War, tomato soup had become a favorite in America's most fashionable hotels. In 1846 the Birch's U.S. Hotel in Washington, D.C., served tomato soup.[11] When Boston's famous Revere House opened in 1847, tomato soup graced the opening dinner. The Cambridge Market Hotel in Massachusetts featured it in the 1850s, as did the famous Parker House, where it initially sold for twelve and a half cents a bowl in 1852. The price increased to fifteen cents per bowl in 1856.[12] The Saint Charles Hotel in New Orleans offered tomato soup as early as 1846.[13] In New York City, tomato soup was served at the Irving House by 1852 and the Dey Street House in 1860.[14] In Chicago's Matteson Hotel, tomato soup was served by 1852.[15] After the Civil War tomato purée was served in fashionable hotel restaurants, including the Stetson House in Long Branch, New Jersey.[16]

Tomato soup often marked festive occasions, as illustrated by the following survey of menus from Boston dated 1856 to 1859. When the Saint Andrew's Society met at the Masonic Temple, tomato soup was served.[17] When Robert Burns's birthday was celebrated by the Burns Club in Boston, tomato soup was on the menu.[18] Tomato soup appeared at dinners held for the anniversary of the birth of Daniel Webster, the New England Society for the Promotion of Manufactures and Mechanic Arts, and the American Insurance Company, "in Honor of the Election of Hon. A. P. Banks, for Speaker of the National House of Representatives," for the city government of Salem, and "in Honor of Turkish Rear Admiral Mehemet Pasha and Suite."[19] When the "Annual Dinner to the Independent Corps of Cadets" was served at Parker's Restaurant in Boston in 1858, the guests consumed tomato soup.[20]

America's most famous chefs emphasized soup, and many of their soups had tomato connections. Felix Déliée, who had served as chef at the prestigious Union, Manhattan, and New York clubs, recommended soup at the beginning of every meal, and he offered menus for 365 meals with a different soup served each day. His *Franco-American Cookery Book* featured many soup recipes with tomatoes, including Sago and Tomato Soup, Rice and Tomato Soup, Nouilles and Tomato Soup, Tapioca and Tomato Soup, and Chicken and Tomato Soup.[21] Thomas J. Murrey, a professional caterer at the Continental Hotel in Philadelphia and the Astor House in New York, believed that the selection of the right soup for the right occasion presented an excellent opportunity for the cook to display good taste and judgment. Arranging and harmonizing a bill of fare was an art form, Murrey believed, and soup was the pivot upon which the meal's harmony depended. His pocket-sized book *Fifty Soups* (New York, 1884) included only one recipe for tomato soup, but many of his other soups had tomatoes as an ingredient.[22] The famous chef Charles Ranhofer also published many soup recipes with tomatoes in his monumental work *The Epicurean* (New York, 1894). Ranhofer, born in Alsace, immigrated to America in 1860. He was the chef at New York's fashionable Delmonico's restaurant from 1862 to 1894. Ranhofer recommended serving two soups—a clear soup and a cream soup—at each meal. His gourmet guide to cookery included more than two hundred soup recipes, such as Consommé à l'Andalouse, Chartreuse Soup, Chicken Okra Soup, Clam Chowder, and Gumbo with Hard Crabs, Creole Style.[23] Oscar Tschirky, maître d'hôtel at the Waldorf in New York, published almost 150 soup recipes, including Tomato Soup, Tomato Cream Soup, Corn and Tomato Soup, Rice and Tomato Soup, Tapioca Soup with Tomatoes, Tomato Soup, and Vermicelli Soup with Tomato Purée. In addition, many other of his soups included tomatoes, such as his American Soup, Andalusian Soup, Crab Soup, and Neopolitan Soup.[24]

NEW DIRECTIONS IN TOMATO SOUPS

New types of tomato soup emerged at the close of the nineteenth century and the beginning of the twentieth: cream of tomato or tomato bisque, tomato bouillon, tomato consommé, and the combination of tomato soup with other major ingredients.

Milk or cream had become an ingredient in tomato soup during the early 1880s. These recipes subsequently were called cream of tomato soup as well as tomato bisque.[25] The first tomato recipe using the word *bisque* originated in New Orleans. Lafcadio Hearn, author of *La Cuisine Creole* (New York, 1885), featured a recipe for Crayfish Bisque à la Créole, which included as ingredients bread, garlic, onions, ham, a turnip, a carrot, flour, water, crayfish, rice, spices and "a few tomatoes," but surprisingly no milk. This may have been a mistake. The following year, *Good Housekeeping* published a similar recipe for Crab and Tomato Bisque, which included milk.[26] Oscar Tschirky of the Waldorf published the first recipe titled Tomato Bisque, similar to the previously published recipes for cream of tomato soup. It combined tomatoes and milk with cornstarch and soda.[27] Janet M. Hill's recipe was slightly more complex, with celery, onions, and flour as ingredients.[28] Tomato bisques are still found in modern cookbooks.[29]

Not everyone preferred milk in soup. Some people were lactose intolerant, and others were lacto-vegetarians, who on principle refused to consume milk. For those not wanting milk, some tomato soup recipes contained peanut milk as a substitute.[30]

The word *bouillon* derived from the French word *bouillir,* meaning to boil. In France bouillons were produced by slowly simmering bones, aromatics, and usually some flesh, frequently augmented by herbs, vegetables, and cereals. Bouillons were considered healthful. American tomato bouillon recipes originated in New Orleans. Celestine Eustis published one in her *Cooking in Old Creole Days* (New York, 1904) that she called Bouillon à la James Madison. According to Eustis it had been developed by an African-American cook who had been born to slaves owned by President James Madison's family. This recipe started with ham bones, simmered with carrots, onions, celery, and a can of tomatoes. The following day, this mixture was strained into a large bowl. The grease was skimmed off, and diced rump beef, eggs, celery, salt, and pepper were added. This was again strained through cloth. When wanted, it was warmed up and infused with celery, chervil, and "a glass of good sherry."[31] Janet M. Hill published four different recipes for making tomato bouillon. Her first, published in 1910, was similar in content to Eustis's, but Hill's recipe retained the tomato pulp, for the soup "should be of a bright red color." Her second recipe, Clear Tomato Bouillon, published six years later, sieved out the pulp. In the same year, her third recipe included oysters. Her last recipe, published in 1917, used both a can of tomatoes and tomato ketchup as the basic ingredients. She sieved out all the vegetables into a napkin. Ever practical, Hill reported that the residue in the napkin could be "pressed through a fine sieve and used in a dish of rice or macaroni."[32] In 1924 Ida Bailey Allen offered two variations on the tomato bouillon theme. The first one made a Quick Tomato Bouillon, which used extract or bouillon cubes instead of stock. She then gave the preparer a choice: either put in tomatoes or spice up the soup with two thirds of a cup of chili sauce. Her second recipe, Jellied Tomato Cream Bouillon, used tomato aspic and whipped cream.[33] Tomato bouillon recipes still appear in cookbooks.[34]

About the same time that tomato bisque and bouillon recipes were published, tomato consommé first appeared. The concept of a consommé dated back to at least the Italian Renaissance, but it was popularized by French chefs in the eighteenth century. Recipes for consommés had been first published in the United States during the 1830s. A consommé came to be defined as a rich, clear broth, which was boiled down, skimmed, and strained. Its base was meat or fish, but not bones. During the nineteenth century, the term was also applied to vegetables. Tomato consommé was an inevitable consequence of applying the process of making a strong, clear broth to tomatoes.

A different approach to tomato consommé was developed in Sarah Tyson Rorer's Consommé with Tomato Blocks. Fannie Merritt Farmer may well have seen this recipe, for three years later she published a similar one for Tomato and Clam Consommé in her *What to Have for Dinner* (New York, 1905). Janet M. Hill published the first recipe for Consommé

Tomatee in the *Boston Cooking-School Magazine* in 1909. These recipes do not produce a clear soup. They begin as a consommé but are then supplemented by tomatoes and other ingredients.[35]

The final major direction was to combine tomato soup with other basic foods. For instance, the *Presbyterian Cook Book* (Dayton, Ohio, 1873) recipe for Corn and Tomato Soup was simply the two base components. Maria Parloa's Green-Corn-and-Tomato Soup recipe was more sophisticated. It used butter, flour, and flavorings. Fannie Merritt Farmer's Huntington Soup was similar, but she spiced it up with onions and cayenne pepper.[36] Others combined tomatoes with sago, rice, tapioca, chicken, peanuts, celery, barley, and spinach.[37]

TOMATOES IN SOUP

Before the first tomato soup recipe was published, tomatoes were used as ingredients in soups with alternate names. The first located recipes calling for tomatoes in soups were published in Mary Randolph's *Virginia Housewife* (Washington, 1824). Randolph featured tomatoes in recipes for veal, barley, and okra soups.[38] It is also surprising to note that Randolph published the first known recipe for gazpacho. How the recipe for this traditional Spanish soup, whose name is from an Arabic word meaning literally "soaked bread," could first be published in the United States relates to who wrote cookbooks in Spain and Randolph's relatives. In Spain gazpacho was a considered a peasant soup. Consequently, recipes for it were not published in early Spanish cookbooks, which were written mainly for the upper middle class. As culinary historian Karen Hess has noted, Mary Randolph probably acquired her Spanish recipes from her sister, Harriet Randolph Hackley, who had lived in Cádiz, Spain. It is also interesting to note that the second and third known published recipes for gazpacho were also not published in Spain. *Novisimo arte de cocina* (Philadelphia, 1845), the first located Spanish-language cookbook published in the United States, featured eighty-eight tomato recipes, including two for gazpacho. This cookbook was printed on a stereotype press for a client in Mexico and was probably not distributed in the United States.[39] It had little influence on mainstream American cookery. While several cookbook authors published similar recipes under the name of Andalusian soup, the term *gazpacho* died out in America until the late twentieth century.

OKRA AND GUMBO

Okra (*Hibiscus esculentus*) originated in Africa. The word is thought to have derived from West African *nkru-ma*. Presumably, slave traders or perhaps the slaves themselves brought okra to the Caribbean, where it was cultivated by 1707.[40] Thomas Jefferson reported that okra was grown in Virginia as early as 1781.[41] Recipes for okra soup had appeared in several cookery manuscripts; Mary Randolph published the first recipe for okra soup.[42] It quickly became a common recipe in cookbooks. Many okra soup recipes contained tomatoes as an ingredient. For instance, Eliza Leslie's *Directions for Cookery* (Philadelphia, 1837) contained an okra soup recipe using bacon, tomatoes, and okra, with lima beans as optional.[43] Lettice Bryan's *Kentucky Housewife* (Cincinnati, 1841) included an okra and tomato soup.[44] An extensive recipe for Okra Soup appeared in Phineas Thornton's *Southern Gardener* (Newark, New Jersey, 1845), also using tomatoes as an ingredient.[45]

Similar to okra soups were the gumbos of New Orleans. Unfortunately, little evidence has surfaced about their early development. At first, gumbos contained okra and filé (dried sassafras leaves), a seasoning thought to have originated with the Choctaw Indians in Louisiana. Creole cooks saw the "possibilities of exquisite and delicious combinations in making Gumbo," reported the editor of *The Picayune Creole Cook Book* (New Orleans, 1901). In addition to okra, cooks constructed gumbos with many other principal ingredients, such as chicken, turkey, squirrel, rabbit, crabs, oysters, shrimp, or even cabbage.[46] The main common element at this stage was the filé seasoning, while tomatoes were optional.

Gumbos migrated quickly throughout America. Early gumbo recipes appeared in agricultural publications. The first known recipe was published in the *American Farmer* in 1830. It was quickly reprinted in other agricultural publications.[47] During the following year, Judge Jesse Buel, future editor of the Albany *Cultivator,* reported that "Gombo" was a very celebrated dish "prepared in those countries" where okra was "grown, by mixing with the green prods, ripe tomatos, and onions; chopped fine, to which are added pepper and salt, and whole stewed." The first gumbo recipe published in an American cookery book appeared in an edition of Eliza Leslie's *Directions for Cooking* (Philadelphia, 1838). She identified it as "a favourite New Orleans dish." Her recipe consisted of okra and tomatoes, but it did not include filé.[48]

Gumbo delighted many admirers. Will Coleman, the publisher of Lafcadio Hearn's *La Cuisine Creole,* described gumbo as the "great dish of New Orleans." According to Coleman, there was "no dish which at the same time so tickles the palate, satisfies the appetite, furnishes the body nutriment sufficient to carry on the physical requirements, and costs so little as a Creole gombo." It was a dinner in itself, the "*pièce de résistance,* entrement, and vegetables in one. Healthy, not heating to the stomach and easy of digestion, it should grace every meal." Most major restaurants in New Orleans presented on their menus three soups and five kinds of gumbo, many of which contained tomatoes as an ingredient.[49]

BOUILLABAISSE

New Orleans was also famous for its bouillabaisse. Bouillabaisse was the classic Provençal fish stew, which was a speciality of Marseilles. According to John Ayto, an editor of the *English Oxford Dictionary,* the French word *bouillabaisse* derived from the Provençal *bouiabaisso,* "to boil and settle." Purportedly, this meant that the cook should boil the stew only for a short period. Bouillabaisse was a fisherman's soup based on trash fish from the Mediterranean, particularly the rascasse or scorpion fish (*Scorpaena scorfa*), but also the sea bass, bonito, conger eel, and other fish and seafood. During the nineteenth century, bouillabaisse became the rage. Cosmopolitan crowds sought out the bouillabaisse served at Marseilles's Restaurant des Phoceens, Basso's outdoor terrace restaurant, and Pascal's *salles.*[50] It became popular in French high cuisine. Recipes for bouillabaisse appeared in several British cookbooks in the 1850s. American cookbook writer Pierre Blot admitted that the real bouillabaisse was made in Marseilles. Imitations, claimed Blot, were made in Bordeaux, other parts of France, and throughout Europe, but they were "very inferior to the real one." However, he offered a recipe based on the fish that could be procured in the United States. He insisted that bouillabaisse was served as a soup.[51]

Modern food writers, such as André L. Simon, claimed that bouillabaisse was "the fishiest of fish soups, with saffron added, but a perfect *bouillabaisse* could not be made except on the shores of the Mediterranean."[52] Perhaps, but, according to New Orleans lore, none other than the famous British author William Makepeace Thackeray believed that "in New Orleans you can eat a Bouillabaisse, the like of which was never eaten in Marseilles or Paris."[53] It is surprising that such a quote emanated from an individual who never visited New Orleans, but Thackeray was indeed an admirer of bouillabaisse. In 1855 he published a "Ballad to Bouillabaisse," which included these immortal lines:

> This Bouillabaisse a noble dish is—
> A sort of soup, or broth, or brew,
> Or hotchpotch of all sorts of fishes,
> That Greenwich never could outdo;
> Green herbs, red peppers, mussels, saffron,
> Soles, onions, garlic, roach and dace.[54]

Bouillabaisse and other soups were firmly ensconced in fashionable restaurants and clubs. Felix Déliée, who had served as chef at New York clubs, prepared bouillabaisse.[55] Thomas J. Murrey, a professional caterer at several hotel restaurants, developed a nontraditional recipe for "Bouille-abaisse" that was based on cod and Shrewsbury ketchup.[56] Charles Ranhofer's bouillabaisse was founded "on red snapper, one-half pound of lobster, one-half pound of perch, one-half pound of sea bass, one-half pound of blackfish, one-half pound of sheepshead, one-half pound of cod, one-half pound of mackerel" and two medium-size tomatoes. American boullabaisse recipes proliferated in the twentieth century, many incorporating tomatoes as an ingredient.[57]

CHOWDER

Whether chowders were introduced into New England by French, Nova Scotian, or British fishermen is undocumented, but chowders were important dishes by the beginning of the eighteenth century in America. The first known American recipe for chowder was published in Boston during 1751.[58] Chowders were quite distinct from broths and soups. Chowders, originally stews, were composed of fish, seafood, and vegetables in various proportions. The object was to prepare a thick, highly seasoned compound without reducing the ingredients to the consistency of a purée. When Hannah Glasse's *Art of Cookery* was first published in America, in 1805, it included the same Chouder recipe that appeared in the 1768 British edition.[59] The recipe was adapted by Mary Randolph, who published it in her *Virginia Housewife* (Washington, 1824). Subsequently, similar recipes appeared in other American cookery books.[60] For instance, J. Chadwick's *Home Cookery* (Boston, 1853) offered recipes for Clam Chowder and Quahog Chowder.[61] As the liquid content of chowders increased, their popularity expanded.

In *The Frugal Housewife* (Boston, 1829), Lydia Maria Child averred that "the best sort of catsup is made from tomatoes" and that a cup of ketchup added to chowder made the dish "very excellent." This was the first known inclusion of tomatoes in chowder.[62] Subsequent cookbook writers also recommended using tomato ketchup to thicken soups and chowders.[63]

Others built on these culinary traditions. In 1867 New York Cooking School director Pierre Blot reported that chowder was made a hundred different ways, but that boatmen were the real experts in making it. He acquired his recipe from the "most experienced chowder-men of the Harlem River." It included layers of potatoes, crackers, clams, and tomatoes but no milk.[64]

Chowder recipes with tomatoes were identified as coming from a variety of different geographical locations. Maria Parloa's recipe for Danbury Chowder incorporated one pint of canned tomatoes or one quart of fresh tomatoes. Boston Cooking School principal Fannie Merritt Farmer's recipe for Connecticut Chowder and Thomas J. Murrey's and Charles Ranhofer's recipes for clam chowder substituted tomatoes for milk.[65] Other recipes included both the tomatoes and the milk, such as Sarah Tyson Rorer's recipes for clam and fish chowder. The *Good Housekeeping Everyday Cook Book* recipe for Rhode Island Chowder included clams, tomatoes, and potatoes, as did Myrtle Reed's recipe for Bar Harbor Clam Chowder.[66]

Several cookbooks included recipes for vegetable chowder or vegetable clam chowder, which featured tomatoes. *Larkin Housewives' Cook Book: Good Things to Eat and How to Prepare Them* (Buffalo, 1915) included a recipe for Vegetable Chowder with canned tomatoes. Fourteen years later, *Good Housekeeping's Book of Good Meals* (New York, 1929) published a recipe for Vegetable Clam Chowder that featured tomatoes.[67]

Many chowder recipes were associated with the New York area. Jessup Whitehead published a recipe for Clam Chowder—Coney Island Style in his *Cooking for Profit* (Chicago, 1893). Presumably, this dish originated at Coney Island, a Brooklyn amusement park established in 1844. Whitehead reported that it was popular in restaurants as a lunch dish and that it specifically contained tomatoes. Joseph Vachon's *Book of Economical Soups and Entreés* (Chicago, 1903) gives a recipe for Coney Island Clam Chowder flavored with tomato ketchup.[68] Margaret Fulton included a recipe for Fulton Market Clam Chowder in her *Grand Union Cook Book* (New York, 1902), presumably named after Manhattan's major fish market. Fulton's recipe included a quart of tomatoes. The *Pictorial Review Standard Cook Book* (New York, 1931) published a recipe for New York Clam Chowder with tomatoes.[69]

The first located recipe titled Manhattan Clam Chowder was published in Virginia Elliot and Robert Jones's *Soups and Sauces* (New York, 1934). This recipe substituted tomatoes for milk.[70] The name Manhattan Clam Chowder caught on, but it was not necessarily associated with Manhattan. Ann Roe Robbins in her *100 Summer and Winter Soups* (New York, 1943) reported tasting Manhattan Clam Chowder at the Greyhound Inn on the shore road to Atlantic City. Robbins, a soup expert, insisted that this was one of the best meals she ever had.[71] Most subsequent authors of chowder offered two recipes, usually one with and one without tomatoes.

The debate as to whether real clam chowder should contain tomatoes was exaggerated by Representative Cleveland Sleeper, who introduced a bill in the Maine legislature in the late 1930s to "make it an illegal as well as a culinary offense to introduce tomatoes to clam chowder." Eleanor Early picked up Sleeper's sentiments, remarking in her *New England Sampler* (Boston, 1940) that there was "a terrible pink mixture (with tomatoes in it, and herbs) called Manhattan Clam Chowder, that is only a vegetable soup, and not to be confused with New England Clam Chowder, nor spoken of in the same breath." She continued by claiming that tomatoes and clams had "no more affinity than ice cream and horse radish." She believed that it was a "sacrilege to wed bivalves and bay leaves, and only a degraded cook would do such a thing." She equated Manhattan clam chowder with a "thin minestrone, or dish water, and fit only for foreigners." She proclaimed that traditional Boston clam chowder was rich and creamy and did not include tomatoes. As historian Richard J. Hooker pointed out in reflecting on this mock debate about the traditional ingredients of chowder, early Americans would "have been astounded to be told that either milk or tomatoes could be used."[72]

Within a few years of the outbreak of the clam chowder war, chef Louis P. DeGouy, author of *The Soup Book* (New York, 1949), concluded that chowder was one of those subjects, like politics and religion, that could never be discussed lightly. In his book, he offered ninety-eight chowder recipes. His recipe for New England Clam Chowder included tomatoes. DeGouy proclaimed that his two recipes for Manhattan-manner Clam Chowder originated in "Gloucester, Swampscott, Nahant, Cohasset, Scituate, all around the Cape and up and down Narragansett Bay from Point to Providence." One Manhattan-manner recipe included tomatoes; the other did not. His recipe for "Pacific Coast manner" Clam Chowder used tomato sauce as an ingredient.[73]

DeGouy presented many other chowder recipes with tomatoes as an ingredient. His Shrimp and Tomato Chowder was reportedly developed at the Hôtel des Roches Noires in Trouville, France. His Tomato Chowder recipe was presented in the "Beauvilliers manner."[74] It was akin to the previously mentioned vegetable chowder. Recipes titled Tomato Chowder circulated beginning in the early twentieth century. Joseph Stoltz, chef de cuisine of Hotel Ponce de Leon in St. Augustine, Florida, was famous for his Tomato Chowder Virginienne consisting of tomatoes, potatoes, onions, peppers, okra, white broth, oysters, and seasonings. Frances Troy Northcross, the "Home Counselor" of the *Washington Times-Herald* and the creator of *Radio Recipes*, included another Tomato Chowder recipe in her *250 Delicious Soups* (Chicago, 1940), a booklet in a series sponsored by the Culinary Arts Institute in Chicago.[75]

Whatever its real origin or ingredients, Manhattan clam chowder is not always appreciated. It has been found to be "rather bland compared to the New England version, and rarely made with the care required." Culinary impresario James Beard once referred to it as a horrendous soup, "which resembles a vegetable soup that accidently had some clams dumped into it." Despite such disparagement, Manhattan clam chowder recipes occasionally appear in cookbooks and on restaurant menus. Today, almost by consensus, it excludes potatoes and milk and includes tomatoes.[76]

TOMATO SOUP'S ETHNIC COUSINS

Tomatoes were entrenched in Aztec cookery well before the arrival of the Spanish conquistadors in 1519. As was previously noted, the Spanish introduced tomatoes into what is today the United States. Their settlements in St. Augustine, Florida, and Santa Fe, New Mexico, preceded English settlements in Jamestown, Virginia, and Plymouth, Massachusetts. Spain also colonized Texas and California. Spanish settlements were composed of peoples from Mexico, some of whom were of Spanish descent, some Amerindian, and some mestizos. Mexico declared its independence from Spain in 1813. When the United States annexed Texas, conquered California and the American Southwest, and made the Gadsden Purchase, it acquired a large number of Spanish-speaking citizens. The culinary traditions of these new peoples were diverse, but they differed significantly from those of Anglo-Americans. Mexican-American traditions survived and thrived in the southwestern part of the United States. Although Mexican cookbooks had been published in the United States, culinary traditions from Mexico did not influence American cookery until the 1840s. The first known "Mexican" recipe appeared in a mainstream American cookbook in 1876. Cookbooks occasionally published other Mexican recipes during the remainder of the century. An early recipe for "salza" published in an American cookbook reported that it could be used as a sauce for meat or an ingredient for soup.[77]

In 1898, a Spanish-language cookbook, *El cocinero español*, was published in San Francisco. Its author, Encarnación Pinedo, came from a prominent *californio* family, and her recipes represented sophisticated Mexican cooking in nineteenth-century California. She was not amused with English cooking traditions, for she reported that there was "not a single English cook who knows how to cook well, and the food and style of seasoning is the most insipid and tasteless as can be imagined." She did include several soup recipes, including the traditional *albóndigas* or meatball soup, which frequently features tomatoes as an ingredient.[78]

May E. Southworth's *One Hundred and One Mexican Dishes* (San Francisco, 1906)—one of the first English-language Mexican cookbooks—also included many recipes with tomatoes as ingredients, among them one for *Cordero*—a lamb soup.[79] Victor Hirtzler, the chef of the Hotel St. Francis in San Francisco around the turn of the century, relished soups. His cookbook included recipes for 111 consommés, thirty-seven cream soups, and 151 potages and thick soups. Tomatoes permeated the recipes. His tomato soups included Onion and Tomato, Tomato with Rice, Tomato Purée, and Tomato Broth, as well as a special recipe titled Potage Mexicaine, which mixed tomato purée with one pint of well-seasoned consommé tapioca.[80]

Other Hispanic traditions have been brought to the United States through the American conquest of Puerto Rico and through great migrations from Cuba and other Spanish-speaking countries in the Caribbean, Central America, and South America. The tomato plays an important role in these culinary traditions, including serving as an ingredient in many soups. These immigrant cuisines have greatly enriched mainstream cookery.

Immigrants also came from poverty-stricken regions of southern Italy, where tomato culture was well established. While the tomato had been introduced and had been used for culinary purposes in Italy in the mid-sixteenth century, it did not become the predominant ingredient in southern Italian cookery until the late eighteenth century. One Italian observer reported that it took the famines of 1745, 1771, and 1774 to induce southern Italians to adopt the tomato. Tomatoes were grown in northern Italy by the late eighteenth century, but evidently they did not become an important part of northern Italian cuisine until after the Italian unification in 1861.[81] During the 1880s Italian immigrants began arriving on the American shore. Immigration increased during the period from 1900 to 1930. For many immigrants, evening meals consisted solely of soup, bread, cheese, and fruit. According to subsequent generations, Italian-Americans had a saying that soup did seven things: it relieved your hunger, quenched your thirst, filled your stomach, cleaned your teeth, made you sleep, helped you digest, and colored your cheeks. As tomatoes were inexpensive, they found their way into most immigrant foods. Many Italian immigrants became market gardeners near major cities. According to one report in 1905, "There is not a single one of the cities of this country yet reached by the Italians where there is available market land near by that is not now receiving vegetables and fruits as the produce of Italian labor." Market gardeners invariably grew tomatoes. Other Italians took jobs in the sale and distribution of food, particularly in the wholesale and retailing of fruits and vegetables. Still others opened restaurants, which served tomato-based soups.[82]

By the mid-nineteenth century, Italian foods and soups had already entered into American cookery. Macaroni and vermicelli soups with tomatoes, for instance, commonly appeared in mainstream cookbooks.[83] English cookbooks with "Italian" recipe sections were published in America as early as the 1850s, although the first cookbook authored by an American with an Italian recipe section did not appear until the 1890s. The first cookbook solely focusing on Italian cookery was published in the United States beginning in 1912.[84] Turn-of-the-century immigrants brought different culinary traditions, and in time, introduced them to mainstream Americans, including tomato-based soups, such as minestrone.[85]

One immigrant from Sicily was Vincent Taormina. In 1905 he launched a small importing business in New Orleans, bringing the foods of Italy to the United States. By 1925 Taormina had merged with his cousin Giuseppe Uddo to form the Uddo & Taormina Corporation. Taormina's son established an importing business in New York City, which was merged with Uddo & Taormina in 1927 to form the Progresso Italian Food Corporation. When the United States entered World War II, all trade with Italy was cut off. Progresso began manufacturing minestrone soup in Vineland, New Jersey. Soon other soups, including tomato soup, were added to its list of products.

Immigrants from Eastern Europe brought *borshch*, which was originally a soup from Ukraine based on cow parsnips, a plant belonging to the carrot family. Like many other foods, *borshch* evolved: beetroot became its defining ingredient. The first located American recipe for it—Borsch, or Polish Soup—was published in 1895. Today, hundreds of different types of *borshch* are available in Ukraine alone. As *borshch* was adopted by Polish and Russian cooks, numerous varieties emerged. In America *borshch* was initially an ethnic food particularly associated with Jewish immigrants, who called it by its Yiddish name, *borscht*. So bound were Jewish immigrants with borscht that during the 1930s Jewish-owned resorts in Pennsylvania and the New York Catskills were referred to as the Borscht Belt. While many variations thrived, *borshch* became codified in America as Russian, Ukrainian, or Polish. Some versions, such as Moscow *borshch*, contained tomatoes, which had been popularized in Russia by Andrej Bolotov. His article "On Love-Apples," published in 1784, described the attractiveness of tomatoes and refuted the belief that they were poisonous. Tomatoes remained a rarity in Russia, except in southern areas. By the late nineteenth century, tomatoes were grown extensively around the Black Sea. They were used culinarily to flavor *shchi*, a souplike peasant stew. During the twentieth century, tomatoes became a component in *shchi* and added to the sourness of the soup. While references to *borshch* appeared as early as the 1820s, recipes for borscht were not published in mainstream cookbooks until the twentieth century.[86]

By the mid-twentieth century, tomato soup reigned as America's best-loved soup. Most cookbooks featured several recipes for it in all its diversity. In addition, many other soups included tomatoes as an ingredient. Ethnic soups with tomatoes were not yet common outside particular regions, but they were poised to break out of their culinary ghettos and enter the mainstream.

CHARITABLE AND CORPORATE TOMATO SOUP

Charitable and community cookbooks emerged after the Civil War. These works were usually written by nonprofessionals and were intended to generate income for a particular community charity or religious group.[87]

The *Presbyterian Cook Book* (Dayton, 1873) included a recipe for Corn and Tomato Soup and one for Macaroni Soup with tomato ketchup as an ingredient.[88] The *Gulf City Cook Book* (Dayton, 1878), compiled by the "Ladies of the Saint Francis Street Methodist Episcopal Church, South," in Mobile, Alabama, featured two tomato soup recipes and tomatoes as an ingredient in recipes for Okra Soup, Vegetable Soup, Cubion, Okra Gumbo and Oyster Gumbo, and Fish Chowder.[89] The cookbook compiled by the Saint Agnes Guild of the Church of the Epiphany titled *Par Excellence, Manual of Cookery* (Chicago, 1888) offered three tomato soup recipes.[90] By the end of the nineteenth century, tomato soup recipes could be found in almost all charitable cookbooks. This suggests that by this date tomato soup had become a part of America's national culinary repertoire. Then, as now, tomatoes were the most common vegetable grown in home gardens. Gardeners faced the same problem as did commercial tomato growers: what to do with the abundant tomatoes during the high season. Canning them for use in soup was a major solution.

Frequently recipes used the names of commercial products, and charitable cookbooks also usually included advertisements. *Par Excellence, Manual for Cookery* included a promotion for the Boston-based Huckins' Soups company, which sold tomato soup.[91] *The "Home" Cook Book* from Cinnaminson, New Jersey, published a recipe for making Cream of Tomato Soup that included a can of Curtice Brothers concentrated tomatoes as an ingredient. Likewise, *Library Ann's Cook Book*, produced by the staff of the Minneapolis Public Library in 1928, included cans of Campbell's tomato soup in recipes for Italian Spaghetti and Ashville Salad.[92] Such recipes were often developed by the corporation that created the soup. This brings us to a second cookbook trend.

During the 1850s a number of businesses, particularly those producing and selling patent medicines, advertised their services in pamphlets replete with cookery recipes. There was usually no particular relationship between the products or services promoted therein and the recipes. These pamphlets rapidly proliferated during the following decades. Many advertising pamphlets included tomato recipes; some featured tomato soup. For instance, the Ransom family operated a drug emporium in Buffalo and annually published a *Receipt Book*, which interspersed advertisements for medicines and unrelated cooking recipes. Their 1885 pamphlet included a simple recipe for Minnie's Tomato Soup based on stock, a can of tomatoes, and a lemon.[93]

In the late nineteenth century, businesses engaged in manufacturing and selling food-related items fixed upon such cooking pamphlets as a means to sell their products. Unlike the earlier commercial cookbooks, recipes in the food-related promotional category specified the brand-name product being marketed. As the food giants emerged in the early twentieth century, they relied heavily on cookery pamphlets to promote their products in an era without radio or television.

By far the largest producer of recipes for using canned tomato soup was the Campbell Soup Company of Camden, New Jersey. Its cooking pamphlets called for cans of Campbell's soup in such recipes as Stuffed Baked Peppers, Spanish Veal Balls, Bluefish with Tomato Sauce, Scallop Cocktail, Indian Aspic, Chilaly, Green Corn Creole Style, Tomato Aspic, Vol-au-vent of Rice with Beef, and Spaghetti à la Campbell.[94]

▪ CHAPTER 5 ▪

Soupy Sales

Many consumers used canned tomatoes to make soup. One cookbook remarked on the convenience of using canned tomatoes. Canners encouraged consumers to buy their wares to make tomato soup. Some recipes were relatively simple: open the can, sieve the contents, add salt and pepper to taste. Most were more elaborate: open the can, add soda, milk, onion, stock, cloves, bacon, a tomato soup mix, pepper and salt to taste, and then sieve before serving.[1] For the consumer not interested in going through all the bother of making soup almost from scratch, there was another alternative: canned soup.

HUCKINS' SOUPS

The first known commercial tomato soup was created by James H. W. Huckins of Boston, Massachusetts, who began making soup about 1858.[2] He was so proud of his "Improved Tomato Soup" that he applied for, and received, a U.S. patent. According to his application, Huckins boiled fifty pounds of beef shin in fourteen gallons of water for fourteen hours, adding water from time to time as required. He skimmed off the fat and strained the broth from the meat through a fine sieve. This produced seven gallons of stock. He added one and a half bushels of mashed tomatoes, boiled the mixture for another one and a half hours, and strained it to remove the seeds and skins. In another boiler, Huckins combined finely chopped onions, turnips, carrots, and beets with three and a half pounds of butter. When the vegetables were cooked, flour, black pepper, and brown sugar were added to the beef-and-tomato stock, and the mixture was also boiled and strained. This formula produced thirteen gallons of tomato soup. Huckins believed that this composition had inherent preservative qualities preventing decomposition "for a great length of time." The U.S. Patent Office agreed and issued a patent on May 2, 1865.[3]

Sometime after this date, Huckins began canning his soup, making him the first known soup canner. He first advertised his canned tomato soup in 1876. As Huckins's soups were extremely successful, other canners entered the field. By1882 Artemas Ward, a Philadelphia grocer and prolific food writer, reported that "soups of all descriptions are now packed in hermetically sealed cans, and are a very great addition to the comfort of the cook, being prepared for the table almost immediately. There are also many extracts made which simply require the addition of water to make a good basis for soup." These extracts or "dry soups," desiccated packages of ingredients for making soups, were put up by Tyrell & Company of New York.[4]

Huckins's soups received national visibility. The *New York Home Journal* reported that they were "commended by all those who use them, for their delicious flavor, their nutritiousness, their cheapness, and their readiness for immediate use. Their sale is consequently increasing, and their name will soon become a household word throughout the country." The *Boston Saturday Evening Gazette* claimed: "Huckins' Soups have a reputation deservedly their own. Those who try them once will be unwilling to be afterwards without this convenient delicacy." It further reported that "Huckins' Soups have been growing in popularity since their first introduction to the public. This popularity is due to the care in selection of material, the skill used in preparing the same, and the determination of the house never to allow imperfect goods to be placed on the market." The *Boston Daily Advertiser* announced that "Huckins' Soups furnish a delicious meal at a moment's notice. They are prepared with the utmost care from the choicest ingredients, and need no cooking." The *Hartford Courant* stated: "The superior quality and richness of Huckins' Soups have given them a very large and increasing sale in all parts of the world. They have been used for years, and are everywhere commended for their excellence, convenience, and economy." Greenly-Burnham Grocery Company of St. Louis tried for several years to substitute cheaper brands for Huckins' Soup but finally gave up and handled "no other brand of canned soups. We can buy other kinds a little cheaper, but the people won't have them, and we cannot sell them." The *Philadelphia Press* announced that "Huckins' Soups enjoyed a deservedly high reputation," because they were "rich, delicious, and carefully prepared." The *Boston Evening Traveller* found them "convenient, economical and *satisfactory.*" The *Boston Daily Globe* proclaimed that "soup, especially at noon, starts a meal off in good shape, and is especially effective in 'giving an appetite' to hundreds of people. The preparation of other dishes for a dinner takes so much time and space that all costs and housekeepers are glad to have a soup ready at hand which can be prepared at a moments notice. Huckins' Soups fill this want exactly, as they only require to be heated before serving. In quality and richness they are superb. They have been tested in thousands of instances, and give complete satisfaction."[5]

The *Boston Post* announced: "Soup in its perfection is undoubtedly the most healthful and appetizing of dishes, but its preparation when especially desirable, involves a steady fire for hours. Huckins' Soups need merely to be heated before using, and are not only convenient for home consumption or excursion parties, but are prepared from the very choicest materials, in the most tempting form, by a *chef de cuisine* of many years' experience. They come in many varieties, and if you desire the best Soups, be sure that you have only Huckins." The *Brooklyn Eagle* wrote: "It is 'the thing' to serve for lunch one of the Huckins' delicate Soups, which with biscuits, etc., constitutes a mid-day meal of rare excellence." The *Boston Evening Transcript* proclaimed: "There is nothing like a good reputation. When one says 'Huckins' Soups,' everyone knows that something always nice is meant. If any one has not yet tried them, he should do so at once, for they are the best canned soups on the market."[6]

Huckins began advertising his tomato soup nationally during the 1880s. In these advertisements Huckins proclaimed that his soups were rich and well seasoned and "always maintained their excellence." All the consumer needed to do was to heat the can and serve. They were sold in two-quart cans and wholesaled for $3.25 per case of two dozen.[7] Huckins engaged in a variety of promotional gimmicks. He gave away "free" samples of tomato soup, for instance, charging only "20c to help pay express charge (generally double that am't) we will send a *sample* can of Tomato soup." Likewise, Huckins advertised in charitable cookbooks, such as the *Par Excellence, Manual of Cookery* published by the Saint Agnes Guild of the Church of the Epiphany in Chicago in 1888.[8]

FRANCO-AMERICAN FOOD COMPANY

A principal reason for Huckins's extensive promotion blitz during the 1880s was the emergence of a major competitor headed by Alphonse Biardot, who had been a soupmaker and canner in France. When a large shipment of his soup was refused by the consignees in South Africa, Biardot's business failed. His wife sold her jewelry, and the Biardots moved to Greece, where Alphonse Biardot was employed as steward in the royal palace. They were not happy in Athens, and about 1880 Biardot accepted an appointment as representative of an English firm of food preservers, John Moir & Sons, Ltd., which unsuccessfully operated a plant in Wilmington, Delaware. Biardot's management was successful, and he decided to strike out on his own in 1886. Purchasing equipment formerly used by Moir & Sons, Biardot incorporated the Franco-American Food Company in Jersey City, New Jersey. Subsequently, the company also maintained offices in East Orange. Its initial line of products was canned soups: consommé, bouillon, tomato, oxtail, vegetable, chicken, mock turtle, green turtle and julienne.[9]

While the Biardots eventually produced many delicacies, their superb soups received their fondest care: nuances of flavor were achieved by combining the "right ingredients, exact timing for simmering, the correct kinds of stock." Their recipes possessed the right combination of quality, technique, and cleanliness. The Biardots made several important marketing decisions. The first was not to skimp on ingredients. This increased the cost. They acquired Thomas Garrett and other salesmen who could "talk to the trade." The salesmen took the Biardots "with them on selling trips to profit by the Frenchmen's personality and color." The Biardots were so pleased with the cleanliness of the factories that they invited visitors to inspect their facilities. For those arriving from New York by ferry, the Biardots had a coach waiting. For those coming by train, the Biardots maintained a special railroad car. For those unable to come to Jersey City, the company issued booklets illustrating its operations. The Biardots fashioned a visit to their factory into a formal function. Amid much bowing and scraping, visitors were ushered courteously through the kitchens.[10]

One early visitor was Mary Virginia Terhune, a popular cookbook author, who published an account of her tour of the Biardots' factory. While Terhune was strongly opposed to the operations of many canning factories and favored exposing all their iniquities, she was suitably impressed with the Franco-American Food Company's operation. In their larder she found "huge shins of beef; chickens, dressed and whole, fair, plump, and free from the suspicion of taint; calves' heads, white and firm; mutton and veal in prime order; barrels of healthy vegetables; parsley, celery, and other soup-herbs; all the appliances needful for the manufacture of divers kinds of soup in a private family, but on a gigantic scale." On the upper story of the factory, she found six sixty-gallon caldrons filled with "meat cut into small pieces, cracked bones, peeled carrots, onions, turnips, parsley, sweet herbs, a little salt, all covered with water." The contents were gently boiled for eight hours regulated by a thermometer. The huge pots were heated by steam, which kept a constant temperature. The top of the kettle was propped open to allow partial escape to the savory vapor. The superintendent spoke against the evils of fast boiling for soups, saying: "Soup which has been boiled rapidly cannot be cleared. There should be no chance, no risk in such matters. Cookery can be reduced to an exact science. The effect of certain causes must be the same always. To suffer a hard boil—even for the minute when the cook's attention is diverted by other work—is to ruin a pot of soup." From the pots, the soup was drawn through a faucet, and the debris left in the caldron was "strained and pressed into the liquor, then clarified." To make broth, the best parts

1. During the high tomato season, trucks filled the roads in South Jersey. These trucks are lined up waiting to deliver their tomatoes at the Campbell Soup Company factory in Camden in the 1920s. Courtesy of the Campbell Soup Company.

2. Layout for the Campbell Soup Company, Camden, New Jersey. Courtesy of the Campbell Soup Company.

3. Cooks with large wooden ladles stirring soup by hand during the 1930s. This task was fully automated a few years later. Courtesy of the Campbell Soup Company.

4. Workers packing boxes of canned tomato soup by hand during the 1930s. Subsequently, the entire process was automated, and all packing was accomplished by machine. Courtesy of the Campbell Soup Company.

5. Trucks waiting to deliver canned Campbell's soups to stores and wholesalers during the early twentieth century. Courtesy of the Campbell Soup Company.

6. Advertisement for Campbell's Tomato Soup circa 1916. Courtesy of the Campbell Soup Company.

7. The "Beefsteak Tomato" label first used by Anderson & Campbell during the 1870s. Courtesy of the Campbell Soup Company.

8. Campbell's soups were sold all over the world in the twentieth century. This is a photograph of a Campbell soup truck in Paris during the 1920s. Courtesy of the Campbell Soup Company.

9. Truck loaded with tomatoes in peach baskets, which held about a bushel of tomatoes and were open at the top. At the corners were posts, so that the baskets could be packed in crates one atop another without damaging the fruit. Courtesy of the Campbell Soup Company.

10. During the peak tomato season, horsedrawn wagons of every description brought tomatoes to canneries. This photograph was taken about 1900. Courtesy of the Campbell Soup Company.

11. Once tomatoes arrived at the loading dock, they were transported by conveyor belt to the factory. Courtesy of the Campbell Soup Company.

12. A Campbell soup delivery in Norway during the 1920s. Courtesy of the Campbell Soup Company.

13. Tomatoes moving by conveyor belt to the factory in 1905. Courtesy of the Campbell Soup Company.

14. Tomatoes were barged in to the Campbell Soup Company during the 1920s. Courtesy of the Campbell Soup Company.

15. (*left*) The Franco-American Food Company was founded by French emigré Alphonse Biardot. Its main product was gourmet soup. The company published several advertising pamphlets, which told the Franco-American story. After the company was sold to the Campbell Soup Company, its soups were phased out, but the brand name is still used to market spaghetti. Courtesy of the Campbell Soup Company.

16. (*right*) The Joseph Campbell Preserve Company published the first tomato soup cookbook in 1914. This booklet included several recipes for using Campbell's soup in dishes with spaghetti, oysters, clams, fish, chicken, turkey, asparagus, and artichokes. The company changed its name in 1921 to the Campbell Soup Company. Courtesy of the Campbell Soup Company.

of the poultry and meats were boiled, diced, and put into a can. The bones and inferior portions were placed in the caldron. Terhune sampled mutton broth just before it was to be canned and found it "hot and delicious." It was not thickened with flour, a common practice at the time. She described high-quality vegetables and meats used in the canning factory. She proclaimed that the Franco-American Company products had a "deservedly high reputation in hospital and sick-room."[11]

On the lower floor of the soup factory the soup was canned. The Biardots manufactured "charcoal tin" cans in their factory. Charcoal was employed in the process instead of mineral coal; the gasses from it entered into and permeated the fusing metal. Charcoal tin was exceedingly malleable and rarely cracked as cheaper cans did, thus exposing the contents to air. The can was quickly made: "The square plate was fashioned into a cylinder, fitted with top and bottom, and soldered in as few minutes as suffice me to tell of it. *All the soldering is on the outside* of the utensil." After the cans were filled and soldered, they were steamed in a closed caldron until the contents were again at the boiling point, a process that tested the integrity of cans. Next the cans were varnished to prevent rust on the outside, and depending on the contents, some cans were lacquered on the inside. Finally, the cans were hand labeled and wrapped in paper. The Biardots' motto was "Care and cleanliness." Throughout the factory, Terhune found "no evil odors, no dirty corners, no repulsive sights, no hint of uncanny devices for bringing plausible results out of equivocal materials. All is honest and legitimate, corresponding so exactly with the methods and materials intelligent housewives employ for similar ends as to rob 'canning' of disgustful terrors."[12]

Terhune was particularly surprised with "the dexterity and neatness with which the excellent materials used in this establishment are handled, boned, cleaned, trimmed and cut up." According to her report, she frequently visited the factory, which "was full of interest and enjoyment." Her belief was that canned soups were particularly useful for "over-tasked housekeepers." They could be acquired "at moderate prices—good and nutritious—can be had without the trouble of making them; for invalids in boarding-houses to be able to get bouillon and broths of the best quality that have not a suspicion of the 'medicated taste' so abhorrent to the sensitive senses. For the country housekeeper—subject at all times to sudden irruptions of visitors, especially in the summer when the kitchen fire must burn low for part of the day, if workers in the room live through the heated term, there should be infinite comfort in the knowledge that her store-room shelves may be made equal to the fiercest midsummer emergency."[13]

The Biardots engaged in extensive advertising, including the distribution of several expensive pamphlets beginning in the late 1890s. The main thrust of these promotional materials was to assure customers that Franco-American soups were well prepared with the highest quality ingredients and to suggest when to use them. At the time, Franco-American's biggest competitor was not other manufacturers but those who made soup in their own homes. Hence, the company suggested that its soup was for guests who dropped in unexpectedly, for persons who were alone and did not want to prepare a big meal, for those coming home from the theater, and for those engaged in yachting. The promotion asked, "Can anything be more acceptable than a plate of hot soup?"[14] Franco-American trademarked the phrase "French Soups," which undoubtedly increased the product's snob appeal. Its visual image was a representation of a child cook wearing a white jacket and short pants.

As could be expected from the above description, the Biardots were strongly supportive of the pure food movement that gained steam during the latter part of the nineteenth century and the early twentieth. The Franco-American Food Company did not use harmful preservatives, and its manufacturing facilities were as clean as could be. Its soups cost almost twice as much as those of some of its competitors yet were quite popular; they were frequently sold by major retailers and were available in most grocery stores.[15] Other companies that manufactured soup were Richardson & Robbins of Dover, Delaware, who had made tomato soup since 1880, and Libby, McNeill & Libby of Chicago, organized in 1868 by the meat packer Arthur A. Libby. But the Biardots' major competitors were the Camden, New Jersey, soupmakers: the Anderson Preserving Company and the Joseph Campbell Company.[16]

THE ANDERSON PRESERVING COMPANY

Abraham Anderson, born in 1834 in Mount Holly, New Jersey, was apprenticed to a tinsmith. For a time, he worked on tinning roofs in Newark. He made refrigerators until the business failed, then moved to Camden to make tin cans in 1862. He went into business canning oysters in Maryland, which was also unsuccessful. About 1865 he returned to Camden and began packing poultry and Beefsteak tomatoes as well as preserves, apple butter, and mincemeat. By 1868 he was producing fifty thousand cans a year, and his major product was tomatoes.[17]

In 1869 Joseph Campbell of Philadelphia joined Anderson as a partner in the firm.[18] Campbell was a farm boy from Bridgeton, New Jersey. He left the family farm and worked as a purchasing agent of fruits and vegetables for a Philadelphia wholesale concern until 1869. The new partners continued to cultivate the image of the "Celebrated Beefsteak Tomato," which, according to their slogan, was "so large that only one was packed to a can."[19] In 1873 the name was changed to Anderson &

Campbell. They widely publicized Beefsteak tomatoes using the image of a gigantic tomato held upon the shoulders of two men, which was trademarked in 1874.[20] The term was also employed on their Beefsteak Tomato Catsup, which survived for almost one hundred years.[21] In 1876 Anderson & Campbell received a gold medal for its preserving work at the Philadelphia Centennial Exposition.[22] When depression overwhelmed the canned tomato industry, the price for canned tomatoes dropped so far that manufacturers could not afford to can new ones. With hard times upon them, Anderson evidently gave Campbell the option of selling out or buying him out. Much to Anderson's surprise, according to his granddaughter, Campbell bought him out.[23]

Cash-rich Anderson lent money to Henry J. Heinz, who had launched a food packing company in Sharpsburg, near Pittsburgh, in 1869. Heinz expanded the business quickly. Due to this rapid expansion, coupled with a major national economic depression in business in 1875, Heinz went bankrupt. The bankruptcy provisions prevented Heinz from owning another business until he had paid off his creditors, which he did over a period of a decade. When Anderson requested repayment of his loan, Heinz, strapped for cash, offered a white stallion as payment. Anderson accepted and proudly paraded around Camden with Heinz's horse.[24]

Anderson also launched a new canning company in Camden. As the "Beefsteak Tomatoes" logo was retained by Campbell, Anderson obtained a trademark in 1877 for "Boston Market" tomatoes with the "pictorial representation of a four-wheeled timber-wagon, with double team and driver, conveying a very large tomato suspended between the wheels."[25] Boston Market was a variety of tomatoes developed during the 1860s. The fruit of this variety developed early, and it produced an abundant crop of smooth, solid, handsome fruit, which, for canning purposes, was excelled by few if any other available varieties.[26] Anderson canned Boston Market tomatoes and also bottled Boston Market Ketchup.[27]

In 1881 Anderson went into partnership with William G. Knowles, creating Knowles & Anderson, which canned fruits, vegetables, preserves, and jellies. The partnership continued until 1885, when Knowles withdrew and the Anderson Preserving Company was incorporated. Among the product lines was Anderson's Concentrated Soups, which came in "17 varieties," one of which was tomato soup. The company's trademark was an illustration of a monk relishing a plate of soup. It was used on labels and other advertising matter. The Anderson Preserving Company was one of the largest companies in Camden, which was a thriving metropolis at the end of the nineteenth century. The Anderson Preserving Company advertised extensively in magazines and was the first national company to advertise on streetcars. It was also one of the first companies to employ electric advertisements, one of which was installed on the roof of the factory overlooking the Delaware River. Unique at the time, this sign flashed two messages sequentially. In 1899 Anderson also rented space on the Boardwalk in Atlantic City, where he exhibited his soups and offered free samples to the public. Despite his success, Anderson retired and liquidated his business in 1904.[28]

Soup was manufactured by many other canners as well. Snider & Loudon in Cincinnati manufactured cream of tomato soup by 1896. According to one of its advertisements, the recipe had been "prepared by the recipe of the best housekeepers with whole, ripe tomatoes, milk, butter and seasoning, delicately flavored." The soup was ready for immediate use: it only needed to be heated.[29]

THE CAMPBELL SOUP COMPANY

After his split with Anderson in 1876, Joseph Campbell changed the name of Anderson & Campbell to the Joseph Campbell Company. He retained the rights to the "Beefsteak Tomato" label, under which the company canned tomatoes and continued to manufacture Beefsteak Tomato Ketchup.[30] Campbell realized that he needed a financial infusion if the business was to grow faster. He acquired two partners in 1882: Walter Spackman and his son-in-law Arthur Dorrance. Dorrance had inherited wealth from his father, who trafficked in flour and lumber at Bristol, Pennsylvania. After increasing the fortune he had inherited, Dorrance was looking for investment opportunities. When he joined Campbell and Spackman, the company's name was changed to Joseph Campbell & Company. After the death of Spackman in 1891, Dorrance and Campbell incorporated in New Jersey under the name of the Joseph Campbell Preserve Company. The articles of incorporation permitted them to manufacture preserves, jellies, meats, fruits, sauces, vegetables, "and goods of all descriptions." The company grew and flourished, even in the five depression years from 1892 to 1897. Beginning with canned peas and asparagus, it added many other kinds of canned and preserved foods. As president of the company, Joseph Campbell continued to take an active role until his death in 1900.[31]

The Joseph Campbell Preserve Company advertised its canned goods as the "Best in the World." If local grocers did not stock its canned tomatoes, the company supplied samples directly to the consumer and pressured local retailers.[32] At this time soups were of no great concern. Among the two hundred products the company manufactured, the ready-to-serve tomato soup was only one product, although it sold reasonably well.[33] This was about to change due to Arthur Dorrance's nephew John T. Dorrance.

John T. Dorrance, born in Bristol, Pennsylvania, studied chemistry at the Massachusetts Institute of Technology. After receiving his degree in 1895, he studied in Germany at the University of Göttingen, where he received his doctorate after only one year of study. When Dorrance returned to the United States, he had offers to teach at Columbia, Cornell, and Bryn Mawr universities, all of which he declined. In the summer of 1897, he went to work in the laboratory of Campbell's cannery in Camden. Dorrance assumed he would make a good salary, but his uncle paid him just $7.50 a week.[34]

Dorrance made the decision to concentrate on canning soup. As the story goes, while in Europe, Dorrance was astounded at the importance of soup in the German diet. Soup was served as a part of almost every meal, even for breakfast. Dorrance concluded that Americans could consume more soup like Europeans. The soups canned by Huckins and the Franco-American Food Company were expensive, retailing at twenty-five to thirty-five cents for a quart can. These canned soups were ready to serve: they required no mixing or additional water. Joseph Campbell's onetime partner Abraham Anderson, then a major competitor, manufactured seventeen varieties of "concentrated soup." The idea of concentrated soup did have several potential advantages. As water had been removed, the can was smaller, resulting in lower manufacturing costs, reduced freight bills, and less costly labels. Smaller cans also required less space to display in grocery stores and occupied less cupboard space in the home. All the consumer had to do was add water, heat, and serve—a task almost anyone could perform.[35] If Dorrance could succeed in mass-producing quality condensed soups, he could undersell the competition. Within a few months of his arrival at Campbell's, Dorrance had produced five "condensed" soups: tomato, chicken, oxtail, vegetable, and consommé.

Campbell's original tomato soup recipe remains a proprietary secret. However, commercial recipes for tomato soup appeared in print during the early twentieth century. The Canning Trade's recipe for tomato soup consisted of tomato pulp, water, salt, butter, sugar, flour, onions, garlic, bay leaves, and cinnamon. Its recipe for condensed soup included paste and some additional ingredients: beef extract, cornstarch, and mace.[36] Similar recipes were published by Clyde Campbell in 1929, although he also included cayenne pepper, allspice, cloves, parsley, celery seeds, and Worcestershire sauce as added flavoring.[37]

Condensed tomato soup, as it first appeared in a Campbell's can with an orange-and-black label, was first produced late in 1897. As Dorrance had only just arrived during the summer, it is likely that work on condensed soups had commenced well before his return from Germany. It is extremely unlikely that he would have had the time to settle in, develop a formula, convince his uncle about its value, figure out how to produce it, and manufacture it in a matter of a few months. Dorrance may well have been involved in perfecting the product after the initial season.

The now famous red-and-white color combination was based on an idea suggested by Heberton Williams, the company's future treasurer. At Thanksgiving of 1897 he attended a Cornell–University of Pennsylvania football game. Cornell's new uniforms were a handsome red-and-white combination. After the game he suggested these colors for the new Campbell label. The black-and-orange label was judged too somber for Campbell's new condensed soup, and the red-and-white labels were substituted in January 1898.[38] Embossed on the label were two torches, with a double eagle in a circle in the middle. In 1899 the eagle was replaced with the image of the gold medallion Anderson & Campbell had won at the Philadelphia Centennial Exposition in 1876. The uncluttered label with the prominently displayed name of Campbell's Condensed Soup reflected the uniqueness of the product. Creating a simple but attractive label was a fateful decision, according to the packaging expert Thomas Hine, who also asserted that "just as the soups were concentrated to save transportation cost and space on shelf and in the kitchen cabinet, so was the expression of the can."[39]

After manufacturing the soups, Dorrance had to persuade grocers to stock them and housewives to try them.[40] Dorrance sought support from his friend and fellow MIT alumnus Robert Wason, then a junior partner in a Boston wholesaler. To convince Wason to buy his soups, Dorrance agreed to demonstrate them in grocery-store windows in Boston. According to Dorrance, even though his former college friends stood outside and rapped on the windows, his demonstrations were successful, and Wason bought a carload of Campbell's soup.[41] Shortly afterward, the Joseph Campbell Preserve Company decided to advertise its soups nationally, saving Dorrance from the embarrassment of further demonstrations.[42]

A new soup laboratory was equipped in 1899, and Dorrance expanded his work, developing new soups and improving existing ones. Dorrance conducted experiments to determine how best to maintain a uniformity of flavor and taste in the products and to reduce waste caused by can spoilage. Campbell's soup received a boost when the company entered its condensed soup into the canned goods competition at the 1900 Paris Exposition Universelle.[43] So proud were Campbell's executives that they printed the medal on the label of every soup can. The medal remains on the Campbell's label today.

Dorrance continued releasing new condensed soups. Eventually, in 1910 Dorrance ended up with "21 varieties" of soups, including several with tomatoes as ingredients, such as Tomato Okra and Vermicelli Tomato. Why Dorrance selected and promoted the number "21" is unknown. However, it was similar to the number themes of his competitors, such as Abraham Anderson's "17 varieties" and Henry J. Heinz's "57 varieties."

In the year that John T. Dorrance arrived, the Joseph Campbell Preserve Company had lost sixty thousand dollars. One year after the introduction of Dorrance's condensed soups, the company became profitable. According to Dorrance, this financial turnaround resulted in an increase in his salary to nine dollars per week in his second year. By 1900 the total sales of the company had reached $580,000, and Dorrance was made a director and vice president. But his salary was pegged at $12.50 per week. Finally, in his fifth year he received his salary plus 5 percent of the profits from the sale of soups. The soup division of the company expanded, while the nonsoup sectors declined in importance. Most of the other two hundred products that Campbell made were phased out, and the major effort focused on condensed soup. Out of total sales of $900,000 in 1904, soup accounted for $750,000.[44]

Dorrance had some unusual means of maintaining the quality of Campbell's soups. One was for him to spend three months a year as an assistant cook in some of the most famous kitchens in the world, including the Café de Paris and the Waldorf in New York. According to Dorrance, "From those famous chefs I learned all that I know first of the delicate flavoring of soups, and the fact that they made me an honorary member of the Societe de Secours Mutuels et de Retraite des Cuisiniers de Paris is one of my proudest achievements. This way I learned not only correct flavoring but the tastes of the general public for whom our soups are made. It was at Paillards that I cooked an order for the prince of Wales. I knew I was cooking for him, but he didn't. Later in the evening I went into the main restaurant and had my own dinner."[45]

Another way that Dorrance improved the quality of his soups was by hiring Parisian-born chef Louis Charles de Lisle. De Lisle emigrated to the United States in 1874, and became associated with the Franco-American Food Company in 1887. Seven years later he moved to the Curtice Brothers in Rochester, New York.[46] Dorrance hired him in 1902. For the next twenty-seven years, De Lisle played an important role in maintaining the quality of Campbell's soups. When he retired in 1929, Dorrance conferred on him the title of chef emeritus.[47] France honored him with the Croix de Chevalier du Merité, Agricole, for "having contributed to the appreciation of the artistry of French cooking throughout the entire world."

When De Lisle received this award, Irma Rombauer was madly writing what would become the first edition of *The Joy of Cooking,* one of the most financially successful cookbooks of the mid-twentieth century. Rombauer, much taken with canned foods, ruminated that it was regrettable that De Lisle's "distinction could not be made to include all soup manufacturers who have brought to us this good and nutritious product at so low a cost."[48]

As soon as Campbell's soup became successful, other canners jumped on the soup wagon. The Philip J. Ritter Conserve Company manufactured tomato soup in Philadelphia by 1902.[49] In Chicago tomato soup was manufactured by the Diamond Company, and the Libby, McNeill & Libby company produced "Concentrated Soups," which sold for the same price as Campbell's soup. H. G. Love Apple Tomato Soup was advertised in 1904.[50] Burt Olney Canning Company produced several soups in Oneida, New York.[51] So did the T. A. Snider Preserve Company in Cincinnati and the Loudon Canning Company in Cincinnati.[52] Most of these products were rarely advertised and soon disappeared. Some tomato soups survived and thrived. The H. J. Heinz Company of Pittsburgh manufactured ready-to-serve cream of tomato soup beginning in 1897 and subsequently tomato soup as well as many others under private labels, meaning that under contract Heinz produced the soup and glued a non-Heinz label on the outside. Today, Heinz remains the largest producer of private label soup in America.[53]

Initially, Campbell's soup retailed for ten cents a can. They sold it to everyone at the same price, assuring smaller operations that they did not give volume discounts that could undersell them. The profit generated by the Campbell Soup Company was less than a quarter cent per can. For larger stores, it meant that Campbell's soup was a loss leader: stores sold it at a loss to bring in customers who bought other items at a profit.[54] Campbell raised the price to twelve cents in 1925—a price that remained constant for decades.[55]

DEPRESSING TOMATOES

The good times at Campbell did not last forever. In 1928 Campbell confronted the first of several major challenges: its tomato crop failed. This meant that Campbell had to purchase tomatoes on the open market, thus lowering the quality of the tomatoes used in its soups and raising the costs.[56] Just as the company was recovering from the tomato disaster of 1928, John T. Dorrance died and the Depression hit.

Arthur Calbraith Dorrance, twenty years younger than his brother John T., succeeded him as president. Like his older brother, Arthur had also attended MIT, graduating in 1914 with a Bachelor of Science degree in chemical engineering. After graduation, he worked in the Campbell's kitchen for a while, then became assistant soup chef at the Hotel McAlpin. Unfortunately, he overturned a soup pot and was fired, but acquired another similar position at the Ritz Hotel. In 1915 John lent Arthur money to purchase the Franco-American Food Company. Arthur became its secretary in 1916 and was named president in 1917. When the United States entered World War I, Arthur joined the army. Before leaving for France, he sold his shares of the Franco-American Food Company to his brother's wife. It was not until 1921 that the Campbell board of directors purchased the shares of the Franco-American Food Company from John T. Dorrance's wife and merged the two companies.[57]

After serving honorably in France as an artillery captain during World War I, Arthur Dorrance was hired at Groton-Pew Fisheries Company in Gloucester, Massachusetts, where he worked for two years. He then returned to the Campbell Soup Company, becoming assistant general manager in 1923, general manager in 1928, and president when his brother died in 1930. When Arthur took over the Campbell Soup Company, it was in a downward financial spin seriously affected by the Depression. Its sales had declined by 15 percent in 1930 and 23 percent more in 1932. Campbell's soup still had a major advantage: its low price. At twelve cents per can, it could provide a healthy and inexpensive meal. To get this message out, Arthur Dorrance invested heavily in advertising. In 1931 Campbell first advertised on radio. By 1935 Campbell's advertising budget was set at three and a half million dollars. This campaign succeeded, and Campbell overcame the slump.[58]

The Depression had two profound effects on Campbell. The first affected the company's globalization. Campbell had exported its soup abroad for decades. During the early Depression, many countries put up protectionist barriers to help local manufacturers. Canada protected its manufacturers from American competition by increasing the import rate for duties on canned goods in 1930. Campbell responded by organizing a subsidiary in Canada, the Campbell Soup Company Ltd. For similar reasons, Campbell's Soups Limited was organized in the United Kingdom in 1933. Not until the twentieth century did the British develop a taste for tomatoes, but by mid-century they had acquired "a passion for tinned tomato soup." Campbell Soup Company quickly became one of the United Kingdom's greatest suppliers of tomato and other soups.[59]

These first subsidiaries in other countries were followed after World War II by many more. In addition, Campbell has acquired companies in other countries, such as Liebig in France and Erasco in Germany. These subsidiaries and acquisitions have greatly increased Campbell's worldwide sales of soup. According to a recent company press release, "Liebig soups in France and new 'Campbell's Deliciously Good' soups in the United

Kingdom contributed to the increase in worldwide soup sales. In Asia-Pacific, soup sales grew at a double-digit rate due primarily to strong gains in Australia where Campbell continues to build upon its market leadership."[60]

In America, the Depression affected Campbell in a different way. During the 1920s Campbell had controlled over 90 percent of the American soup market. In 1935 Campbell still canned two thirds of all soup canned in America, but its competitors were closing in. The H. J. Heinz Company of Pittsburgh controlled one sixth of the soup market and was gaining on Campbell. The Hormel Company, originally founded by meat packer George A. Hormel of Austin, Minnesota, started making soup in 1932, and by 1935 Hormel was gaining on Heinz. An even greater challenge to Campbell's dominance came from the Phillips Food Company of Cambridge, Maryland, named for Colonel Albanus Phillips. Phillips produced fifteen condensed soups, seven of which sold for five cents per can.[61] To meet these challenges, the Campbell Soup Company increased its advertising budget and diversified its product lines beyond soup.

THE JUICY TOMATO

The Campbell Soup Company's first major diversification was in tomato juice. The drinking of tomato juice was a mid-twentieth-century phenomenon.[62] According to several accounts, tomato juice was the creation of the American-born French chef Louis Perrin. In 1917 he experimentally served tomato juice to his guests at a resort in French Lick Springs, Indiana. Chicago businessmen who spent their vacations in French Lick Springs purportedly spread the word to others about the "tomato juice cocktail in lieu of stronger mixtures."[63] Although canned tomato cocktails were growing more popular by the 1920s, none of the existing products yielded juice with just the right color and flavor.[64]

Tomato juice cocktails were heralded during a Tri-State Packers Convention at Philadelphia's Adelphia Hotel in 1922. A can manufacturer served tomato juice free of charge to each participant in the annual banquet in hopes that canners would pack tomato juice. By this date, tomato juice was touted as a health drink and was served in hospitals. According to Dr. Hugo Friedstein of Chicago's Children's Memorial Hospital, the vitamin content of tomato juice did "marvelous things in cleansing the system." Yet it was not canned commercially.[65]

The reason for the failure of canned tomato juice was that tomato solids settled at the bottom of the can, or the glass the juice was poured into. In 1924 an Indianapolis pediatrician discussed this problem with his friend Ralph Kemp of Frankfort, Indiana. Kemp had majored in agricultural engineering at the University of Wisconsin and at the time worked with his father, John Kemp, operating a canning plant. Intrigued with the challenge, the Kemps began experimenting to find a way to break tomato pulp into minute particles that would float in the juice. Their solution was to use a viscolizer previously employed in the manufacture of ice cream. It required a great deal of adaptation to be used successfully canning tomato juice.[66] After four years of work, the Kemps finally succeeded. In 1928 they applied for a patent and initiated the first national advertising campaign for their tomato juice.[67]

Tomato juice was an instant hit with the American public. The Campbell Soup Company moved into high gear to produce its own tomato juice. Campbell converted part of Camden's Plant No. 2, built during the 1920s, for making tomato soup. The problem now became what tomato variety should be grown for making juice. After experimentation, a tomato was found that met the needs. Campbell released its version of tomato juice in 1931.[68] In 1932 Campbell launched a major marketing drive for its tomato juice, and by 1935 30 percent of Campbell tomatoes went into making that product.[69] By the following year, cookbooks included recipes using Campbell's tomato juice as an ingredient.[70]

Another reason tomato juice was so successful was the end of Prohibition. A cocktail made of tomato juice and vodka was probably first developed at Harry's Bar in Paris by Ferdinand "Pete" Petiot. Petiot moved to New York in 1933 and introduced his new creation. After experimentation, he added Worcestershire sauce and called it a Bloody Mary. Its name was supposedly derived from the British Queen Mary I, who killed many Protestants during her reign in the mid-sixteenth century. Others claim that Mary was Petiot's girlfriend. Whatever the origin of the name, the cocktail quickly conquered America and the world. American writer Ernest Hemingway claimed in a 1941 letter to have personally introduced the Bloody Mary into bars in Hong Kong.[71]

Tomato juice was a natural addition to the Campbell Soup Company, which had been built on the tomato. Its introduction, however, reversed the corporate decision to focus solely on soupmaking. The implications of this trend would not be felt for decades. Shortly after World War II ended, Campbell Soup Company made another logical addition when it purchased V8 juice from Standard Brands. V8 vegetable juice was a blend of eight vegetables along with several flavor enhancers. It had been conceived in 1933 by W. G. Peacock of Evanston, Illinois. Several people worked on the formula. Peacock interested three investors, and the New England Products Company was created. The product was created in 1936 under the name Veg-min. At the first store that sold it, a clerk suggested that they change its name to V-8, which Peacock did. Later the hyphen was removed and the product marketed as V8 Cocktail Vegetable Juice.

Peacock's entire operation was accomplished by hand, and he only had the ability to produce twenty-five cases per day. V8 juice was a success, but he did not have sufficient manufacturing capability to meet the demand. Neither did he have the money to market it properly. In addition, he had other potential products that he wanted to test-market. To raise cash in 1938, Peacock sold the V8 formula to Loudon Packing Company, which had been founded by Charles F. Loudon. In 1891 Loudon went into business with Robert Skinner and for a time manufactured tomato soup. The partnership was dissolved after a few years, and the Loudon Packing Company emerged.

Loudon's company was a major national tomato packer, but during World War II the company faced many problems. It was purchased by Standard Brands in 1943. By the time Campbell purchased V8 juice from Standard Brands in 1948, its annual total sales generated five million dollars.[72] In addition to V8 juice, the Campbell Soup Company eventually acquired many other interests as diverse as Pepperidge Farm and Godiva Chocolates.

During the 1930s the tomato was king at the Campbell Soup Company. From late August to mid-October, all other soup production halted for making tomato soup. During the peak season Campbell's Camden plant produced ten million cans of tomato soup per day. In 1935 tomato soup accounted for almost half of Campbell's total annual soup volume.[73] Subsequently, tomato-related products declined as an overall percentage of the Campbell budget. The tomato was never again Campbell's supreme product as it had been during the 1930s. However, tomato soup remained Campbell's largest-selling soup. In 1970 Campbell's president, William B. Murphy, proudly announced: "We think we know more about the tomato than anyone else in the world."[74]

ADVERTISING TOMATO SOUP

Campbell's sales rose from a half million cans in 1900 to seven million by 1914 and to eighteen million by the early 1920s.[75] Campbell's total budget for all expenses increased from $580,000 in 1900 to almost $6 million in 1912 and to $25 million by 1920. By 1927 sales exceeded $50 million.[76] There were many reasons for this rapid and consistent growth. The Campbell Soup Company had good products and excellent leadership. But, most important, the company knew how to advertise and market its products.[77]

Previously canners sold their goods to regional middlemen, who broke up the lots and sold to local stores. The cans were sold with or without labels. Middlemen purchased the less expensive cans without labels and placed their own on the can. Labels at first were relatively simple, but as the use of brand names increased during the latter part of the nineteenth century, these labels became more attractive. At the same time, the U.S. Patent Office began registering trademarks and slogans. With brand names corporations could advertise nationally, thus generating demand when customers would ask for the brand-name product at local grocery stores. These stores in turn went directly to the manufacturer to procure the goods, thus eliminating the need for middlemen or brokers and reducing the price for the retailer and customer.

The discovery of the importance of promotion created the need for the advertising agency that could influence the customers directly. The pioneer agency in making brand names well known was N. W. Ayer and Son of Philadelphia. Previously, the agency had mainly purchased advertising space for customers in newspapers and magazines. For those customers unable to compose their own copy for the advertisements, Ayer created the position of copy editor. N. W. Ayer and Son developed expertise in establishing brand identity. The agency provided advertising advice for Procter & Gamble Soaps, Burpee Seeds, and the New Jersey firm that manufactured Hires Root Beer, and major retail stores such as Montgomery Ward. The position of the advertising company was fully established when the National Biscuit Company approached Ayer to help create a market for its new product, Uneeda Biscuits. In 1898, the first year that the biscuits were promoted nationally, 120 million packages were sold. This was a staggering achievement, given that the annual sales of Uneeda Biscuits' competitors totaled somewhere around six million. Other manufacturers tried to latch on to this overwhelming success by marketing products with similar names, such as Uwanta Beer and Isagood Soup.[78]

The message of Uneeda Biscuits' national sales campaign was not lost on John T. Dorrance, who convinced his uncle that advertising was crucial. The company's interest in promoting its products sharply increased. Campbell's advertising budget was set at $50,000 in 1901 and reached $135,000 in 1904. The budget steadily rose almost every year subsequently. When sales reached almost six million dollars in 1912, the advertising budget topped half a million. By 1919 the advertising budget was set at $800,000, and the following year, it topped one million.[79]

Early advertising emphasized cards in streetcars. Leonard M. Frailey, secretary for the company at the time, stated that streetcar advertising was selected because "we believe we reach the clientele we are most after, that is, the women. They are frequent riders in streetcars." The company was advised to reach women through "child appeal." Theodore Wiederseim, an advertising agent, was asked for suggestions. He recommended the services of Grace Drayton, who drew plump children. Wiederseim presented the drawings in a portfolio. The concept was accepted, and the company adopted what would become known as the Campbell Kids. The Kids appeared on streetcar advertisements in early 1905 and in magazines in September 1905. At first the advertisements were in black and white. The company introduced advertisements with a red color the following year. Full-color advertising did not commence until the 1920s.[80] From the beginning, the Campbell Kids advertised tomato soup.

Newspaper and streetcar advertising declined as the company found that magazine advertising gave it a much broader exposure. Tomato soup was extensively advertised in such national magazines as the *Saturday Evening Post, Good Housekeeping,* and the *Ladies' Home Journal.*[81] In 1912 an advertisement for Campbell's tomato soup proclaimed that it was "wholesome," had "a mighty 'fetching' quality about it in every sense of the word; and equally for the young and the old. All people and all tastes seem to agree on **Campbell's Tomato Soup**." The advertisement went on to claim: "The epicure approves it. The romping hungry schoolboy craves it and thrives on it. The brain-worker, the nervous dyspeptic, and the man who wants a 'good square meal' all find satisfaction in its tempting flavor and nourishing quality. There's hardly a day in the year when you will not find this perfect soup exactly what you want." The advertisement ended with the rousing question: "Hadn't you better order a dozen *today?*"[82]

Another advertisement reported: "No human chemist ever invented a bracer superior to the delicious recipe put up by nature in the juicy vine-ripened tomatoes used in **Campbell's Tomato Soup**." It continued: "All their valuable tonic and medicinal properties are retained by the Campbell method, while the other nourishing materials which we blend with nature's formula complete a soup as beneficial as it is tempting. Summer is just the time when you need the healthful and appetizing stimulus of this wholesome Campbell 'kind.' Its regular use at this season will do the whole family a world of good."[83]

Campbell's advertising staff was small when compared to that of other national food sellers, such as General Foods or Standard Brands, yet Campbell spent heavily on advertising. When the company ran into difficulty, it spent even more. In 1928 the company spent $2.2 million in magazine advertising alone. The Depression hit in the 1930s, and the company's advertising budget shot up to $3.5 million. Some of those dollars went into radio beginning in 1931. By 1934 the Campbell Soup Company sponsored the George Burns and Gracie Allen radio program. During the 1930s Campbell Soup Company introduced the jingle "M'm! M'm! Good!" When Americans started to turn on television, Campbell's advertising was there, beginning in 1950 with its sponsorship of *Saturday Night Review*. It also sponsored *Lassie* and featured a long list of celebrity endorsers, including Ronald Reagan, Johnny Carson, Jimmy Stewart, Orson Welles, Helen Hayes, and Donna Reed.[84]

Campbell also advertised its products effectively through the regular production of cookery pamphlets and cookbooks emphasizing how its soups could be enhanced or could be used as ingredients to make other dishes. Its 1910 *Campbell's Menu Book* started with the assertion that the success of dinner depends on the selection of the soup. This booklet offered menus "of educational value" for housewives "not conversant with the kinds of soups that may be appropriately and correctly served with certain meats." Its forty-eight pages contained thirteen menus, along with many additional recipes developed by Cornelia C. Bedford, identified as "an eminent authority on culinary topics." These recipes included several ways of serving a variation of Campbell's tomato soup enhanced with whipped cream, hot milk, cheese, noodles, parsley, celery, vermicelli, rice, or flour. Others used tomato soup to make croquettes and jelly. The recipe for Campbell's Soup Tartare creates a sauce for fish, "hamburg steak," codfish balls, or cold meats.[85]

The Joseph Campbell Preserve Company published the first tomato soup booklet in 1914: its base was shaped like a tomato, and its top was a Campbell's tomato soup can. It began with the statement that for years tomato soup had "been enjoying a constantly increasing use as a dinner course, but only of late have its possibilities for other purposes been coming to our attention." The booklet described how tomato soup was manufactured and offered twelve recipes for using Campbell's tomato soup to make dishes such as sauces for spaghetti, oysters, clams, fish, chicken, turkey, asparagus, and artichokes.[86]

The 1916 booklet, titled *Helps for the Hostess,* included recipes for using tomato soup to make such diverse dishes as Spanish Veal Balls, Bluefish with Tomato Sauce, Scallop Cocktail, Chilaly, Indian Aspic, Green Corn Creole Style, and Oriental Roast.[87] Campbell sent these recipes out to newspapers and magazines and encouraged them to reprint them with "Campbell's Tomato Soup" as an ingredient. Although the recipes were frequently reprinted, the Campbell soup identification was occasionally replaced with "condensed soup."

These recipes tended to ramify through the American culinary world. The *Eagle Cook Book* included a recipe for tomato soup on toast.[88] Irma Rombauer's *Joy of Cooking,* one of the most popular American cookbooks in the twentieth century, recommended keeping a can of tomato soup on the emergency shelf. She considered it "delicious diluted with equal parts of milk."[89] Her recipe for Canned Tomato Soup included one can of Campbell's tomato soup.[90] Other cookbooks used Campbell's tomato soup in such diverse recipes as Welsh Rarebit, Rinktumidity, Ham Luncheon Dish, Tomato Cheese Salad, Chili con Carne, and French Dressing.[91]

The *Ladies' Home Journal* published the Campbell Soup Company's recipe for Aspic Salad, which included a can of its condensed tomato soup. This recipe was cut out by Carlton Lake's mother, who passed it on to Carlton's wife, who in turn served it at a dinner where Alice B. Toklas spoke about the cookbook she was then writing. She was using her own recipes but also planned to include some from her friends. She wanted the aspic salad recipe. "It's absolutely perfect. I've never had one that good," said Toklas. Carlton's wife tried "to fade quietly into the background," for the recipe included "embarrassingly low-brow ingredients: Campbell's condensed tomato soup, Philadelphia cream cheese, Miracle Whip salad dressing." In the end she agreed to send it to Toklas in the morning. True to her word, *The Alice B. Toklas Cookbook* included the recipe attributed to "Mrs. Carlton Lake." True to the original recipe, it included "1 can Campbell's Condensed Tomato Soup."[92]

Some recipes promoted by Campbell were slight revisions of tomato recipes previously published. Tomato toast recipes, for instance, had been published since 1844. It was basically stewed tomato and toast. Campbell replaced the stewed tomatoes with condensed tomato soup. This recipe was published in many newspapers, such as the *La Crosse Evening Leader Press*, the *Lawrence Evening Telegram*, the *Lowell Morning Courier Citizen*, the *Fort Wayne Morning Journal Gazette* and the *El Paso Evening Herald*.[93]

A most unusual recipe appeared in *The New England Hotel Women's Cook Book* (Boston, 1933). Titled Tomato Soup Cake, it called for a can of tomato soup, soda, raisins, flour, cinnamon, and cloves.[94] Subsequently, this recipe appeared in cookbooks under the name Love Apple Cake. It is not known whether this recipe was influenced by the Campbell Soup Company's efforts to develop and disseminate recipes or, alternatively, encouraged the Campbell kitchens to develop a tomato soup cake recipe. According to the Campbell staff, this recipe originated from experiments in making steamed fruit-and-nut pudding. Their first recipe for Campbell's Tomato Soup Cake was published in the *New York Times* in 1949 and subsequently appeared in several magazines. It has been published under a variety of names, including Tomato Soup Spice Cake, Holiday Spice Cake, Tomato Surprise Cake, and Halloween Cupcakes, and has three major variations: a layer cake, cupcakes, and a Bundt cake. Based on Campbell's records, this recipe has been published over one hundred times in newspapers, magazines, and cookbooks.[95]

The Campbell Soup Company has regularly continued to publish cookbooks featuring its products. In addition, many recipes were printed on soup can labels and in Campbell's advertisements. It published *Wonderful Ways with Soups* in 1958 and *Cooking with Soup*, four years later; both were instant successes. Over six million copies were sold, and the pamphlets went through several revisions. *Creative Cooking with Soup* followed in 1985. While the company published many other cookbooks, none was promoted as extensively or achieved the same level of consumer awareness.[96] During the 1950s Campbell's personnel demonstrated preparing their recipes on radio and television. More recently, these recipes have appeared online at the Campbell Soup Company Web site.

In 1994 the Campbell Soup Company was inducted into the Marketing Hall of Fame. According to the award givers, "Combined, Americans purchase on average approximately 70 cans of Campbell's every second of every day of the year. It's a phenomenal marketing success."[97]

▪ CHAPTER 6 ▪

Souper Revolutions

When the word *revolution* is mentioned, most people think of highly visible political upheavals, such as events associated with the American, French, or Russian revolutions. These were short in duration but had highly visible long-term consequences. Other changes, albeit less visible, have been underway in the food processing world that have influenced our lives just as dramatically. Nowhere has this been truer than in the industry that processes tomatoes and manufactures tomato soup. During a relatively short period, the tomato soup industry went from labor-intensive mom-and-pop shops to fully automated global businesses. Along with automation came new breeds of tomatoes and new agricultural techniques and machinery. These quiet revolutions have had a major impact specifically on central California and southern New Jersey. For consumers nationally, these changes have meant easily accessible and highly economical canned goods.

THE AUTOMATED TOMATO

Before 1870 all aspects of tomato sowing, growing, harvesting, and processing were accomplished by hand. Farmers cultivated the tomatoes by hand and brought them to canneries in horse-drawn wagons. By hand, tomatoes were skinned, cored, and trimmed. They were boiled in large pots by cooks who stirred and tasted the contents with large wooden ladles. The cans were fashioned by hand, and the seams and lids were soldered on by skilled tinmen. The cans were hand-carried to the fillers, who in turn handed them over to the cappers. The finished cans were labeled by hand and individually packed in cases. The cases were hand-loaded onto wagons and carted to stores for sale. Over a period of about seventy years, this changed to a fully mechanized and automated system. This phenomenal shift was the result of a series of discoveries in two interconnected areas: canmaking and tomato processing.

Initially, all cans were handmade. The body and lids for each can were measured and marked on tin plate. These were cut out by hand with shears. The edges of the metal were butted together and sealed with a heavy ridge of solder. The tops and bottoms were soldered on. Subsequently, the edges of the ends were turned up by means of a mallet and a piece of iron, known as a heading stake. The edges of the body of the can were lapped to facilitate soldering and to make a better seal. Tinkers who turned out sixty of the cap-hole cans in a day were considered master workmen; the average was less. In 1823 a Frenchman named Angilbert devised a system in which holes were punctured in the lid, the food was placed in the can, and the lid was soldered in place. The can was then placed in the water bath. The hole in the lid provided a means to exhaust the can. When the bath was finished, the hole was soldered, and the can was totally sealed.[1] This was a slow, tedious process.

In 1847 the process began to accelerate when a stamping machine for making cans was invented by Allen Taylor. Two years later the pressed top was invented. These innovations made it easier to cut the tin, but the sides and lids still needed to be assembled by hand. A major advance in canmaking was credited to W.H.H. Stevenson, whose machine automatically soldered the ends to the body of the can. It was known as the "floating" process. Another machine was invented to clap the can around a horn and lap the edges together so that the canmaker only needed to apply the solder to the seams. These inventions permitted one man to produce about twelve hundred cans per day.

A side-seam soldering device was added in 1885. Five years later, the first complete system to produce cans using the "merry-go-round" machinery was developed. This device streamlined production and had the capacity of thirty-five thousand cans per day. By 1900 there were two hundred can manufacturers turning out seven hundred million cans per year. The entire canmaking process had become automated, but the cans had defects. Their tops with the holes were still soldered on. Small disks were sealed over the holes with a drop of solder. Automatic capping machines were invented, but cans made in this way had several drawbacks. Each component was assisted by a machine, but most equipment and all soldering was still managed by hand. More important, the heat from the solder carbonized part of the can's contents and formed black specks in the preserved food.[2]

Although the adoption of sanitary cans occurred over a very brief period during the first few years of the twentieth century, it was the culmination of a series of inventions dating back to 1859, when double-seamed cans were first made as powder canisters. A number of years later, European canners applied the process to sealing containers for food. They placed a thick rubber gasket, similar to those used in Mason jars, between the lid and body. This method was demonstrated by a German firm at Chicago's Columbian Exposition in 1893. The rubber ring, however, was cumbersome and costly. Charles M. Ams experimented with lining the edge of the can with the rubber in a solution of benzene or toluene. This invention reduced the amount of rubber used and simplified the sealing process. The Max Ams Machine Company developed a line of canning machines, and the use of the sanitary can spread quickly. Machines cut strips into proper lengths, and a mechanical trimmer produced the finished product. The American Can Company was formed in 1901 and took over almost the entire canmaking business of the country. Its output in 1914 was three billion cans.[3]

Many other inventions improved canning technology. As previously mentioned, it was necessary to boil cans for five or six hours at first. Isaac Solomon's rediscovery of calcium chloride greatly reduced cooking time. Unfortunately, the calcium chloride had to be cleaned off the cans before they could be labeled. A better solution needed to be found to make it possible to heat cans for short periods of time without using calcium chloride. Actually, the solution had been devised earlier by Nicholas Appert, who employed the autoclave, a pressure cooker–like device that heated cans by steam in an enclosed space. This created high temperatures and caused less strain on the cans, as the pressure inside them was equal to the pressure outside. Also, it was not necessary to clean the cans after heating in the autoclave. Unfortunately, autoclaves had a reputation of exploding, and many canners refused to acquire them. In 1871 Baltimore canner A. I. Shriver invented a safe closed kettle for cooking with superheated water or live steam. This invention led to the development of steam retorts, which revolutionized the canning industry.[4]

The first canneries were labor-intensive operations. The boilers were heated by wood, usually cut from forests surrounding the cannery. The water was usually piped in from a nearby river or was carried in by water wagons. Tomatoes were picked into peach baskets supplied by the cannery. These baskets held about a bushel of tomatoes and were open at the top. At the corners were posts, so that the baskets could be packed atop each other without damaging the fruit. During the peak season, wagons loaded with tomatoes filled the roads. When the farmer arrived at the cannery, he was met by a "wagon master," who assigned him a place in the procession. Few canneries were able to use up by nightfall the tomatoes on the wagons that were already standing in line by early morning. Most farmers unhitched their horse, went home for another load, and returned with it before the cannery was ready for the first load. Sometimes twenty-four hours passed between the time a wagonload was delivered to the wagon master and its arrival at the scales, where the wagon, baskets, and driver were weighed together. When the weight was registered, the wagon pulled up to the scalder, where the tomatoes were loaded into iron baskets, which a man with a rope lowered into boiling water. A few moments later, the tomatoes had been scalded and washed.

The baskets were then carried to a long table surrounded by women and children skinning, coring, and removing blemishes and rotting areas, and throwing the tomatoes into a bucket. The skinner was given three or four cents for each full basket. Most women averaged ten buckets per hour. The contents of the buckets were thrown into the "stuffer," where the cans were filled. The cans were then moved by hand to the cappers, who soldered on the lids one at a time. Twenty-four capped cans were then put into iron trays that were stacked atop each other and hoisted by block-and-tackle into large cauldrons, heated by wood. After scalding, the trays were removed and stacked on the factory floor to cool. The following day the cans were carefully inspected, and those not properly processed were removed. A few days later the cans were hand-labeled and packed into cases for shipping.[5]

The only skilled workers in the canning factory were the cappers. They were usually under the control of the "boss capper," who, under contract with the manager of the cannery, furnished his own men. The cannery was at the mercy of the boss capper, who could decide to go on strike at any time. As the tomato season lasted only a few months and tomatoes quickly spoiled after picking, the cappers could easily demand almost anything from the managers. Some cappers made as much as fourteen dollars per day. It comes as no surprise that one of the first pieces of equipment to automate the cannery was the Jones Capper, which worked with six cans at a time. It was not particularly successful, but its adoption was strongly opposed by the cappers, and violence frequently erupted.

J. D. Cox of Bridgeton, New Jersey, developed a more effective capper in 1887. When properly operated, the Cox Capper reduced the expenses of the tomato cannery by a third. It was still hand-operated, but it was a significant shift toward mechanical production. At the same time, a mechanical soldering machine, known as the Little Joker, was invented. It rotated the cans at an angle in a little bath of melted solder to make the outside seams for tops and bottoms. These two machines increased the ability of canneries to make cans fivefold and circumvented strikes, which had erupted two or three times in each factory during the high tomato season. Luddite-like labor strife faced canneries installing the new machinery, but the days of the boss capper were numbered. By the late 1880s unskilled laborers capped ten cans in a single operation using automatic capping devices. The process was improved even more with the introduction of the Climax Capper, the first successful power capping machine, developed by George Colket and George E. Lockwood.[6]

Another important invention was the foot-powered Bucklin Tomato Filler. Early machines were very messy. As much as one fourth of the tomatoes ended up on the floor or on the operators, who were encased in oilskins. Initially, operators were drenched in tomato juice after two hours of operating the machine. More efficient fillers soon appeared on the market. One observer estimated that with the Stevens Tomato Filler, invented by John Stevens of Woodstown, New Jersey, six hours passed before the operator was drenched with tomato juice. These machines were superseded by mechanical gang fillers, which filled six cans at a time and ran under steam power.[7]

Before long every stage of the canning operation had a machine associated with its operation. Devices for scalding, topping, and wiping were introduced, as were power hoists and cranes. About 1884 cans and bottles made in America were first labeled by machine, but there were problems with speed and paste. In 1890 a machine developed by Fred H. Knapp of Adrian, Michigan, overcame the paste problem. Two years later, conveyor belts were added to the machine. They sped up the process tenfold. Wrapping and boxing machines also came into use about this time. Lines of interconnected equipment were developed by equipment manufacturers such as the Sprague Canning Machinery Company. This firm was bought out by the Remington Machine Company of Wilmington, Delaware, which developed the Triumph line of tomato cannery equipment. New machinery released about 1903 provided for the fully automated manufacture of sanitary cans without the need for solder, thus doing away with cappers.[8]

The final frontier of automation was peeling. In 1914 several canners predicted that this would never be automated. A few years later, peeling machines came into use. By the 1920s the process of canning tomatoes was fully automated. From the time tomatoes arrived by truck until the canned goods were shipped out the back door, the tomatoes were never touched by human hands. Subsequent machines sped up the process and made it more efficient.

Soupmaking did not require peeling the tomatoes but did require a few additional machines and processes. During the 1930s the tomatoes went into breakers and then were pumped to "cyclones"—large centrifuges that removed the skins and seeds. The tomato pulp was subsequently cooked. Using giant wooden spoons, cooks blended the pulp with butter, salt, sugar, flour, onions, and spices. From the kettles, the soup was pumped through stainless-steel pipes to the canning machines.[9] Within ten years, this process was fully automated.

THE WELL-BRED TOMATO

In 1909, John T. Dorrance began acquiring 176 acres of prime agricultural land in Cinnaminson, New Jersey, to establish an agricultural research center. To manage the facility, Dorrance hired Harry F. Hall, a professor at the New Hampshire Agricultural College and the president of the American Vegetable Growers Association. In 1912 Hall began conducting extensive field trials on tomatoes. His major task was to breed a new variety with the right qualities for soupmaking. He selected plants from available varieties, then crossbred those he thought had the right balance of acid and sugar, a thick skin, a deep shape, and the right color of red. It usually took at least six generations for a line to breed true.[10]

Successful breeding meant hand pollination. This required, reported Hall, a light and patient touch. The tomato flower's stamens had to be removed carefully with small pointed pliers twenty-four hours before the flower opened. Within the next twenty-four to forty-eight hours, pollen from the other tomato variety was introduced. Hall named one successful tomato variety developed in 1920 the J.T.D., honoring John Thompson Dorrance. It purportedly derived from the Greater Baltimore tomato. It was popular for canning purposes because it ripened from the inside out; growers could look at the tomato's exterior and have a good idea about the interior state of maturity. The major negative characteristic of the J.T.D. was that it was not particularly disease resistant.[11]

Hall continued working to perfect a canning tomato. In 1928 he crossed the J.T.D. with the Marglobe. The Marglobe had been developed by F. J. Pritchard of the U.S. Agriculture Department from a cross in 1917 between the Marvel and the Globe. The Marvel (*Merveille des Marches*) was of French origin and had been imported into the United States in 1901. Pritchard developed the 1917 cross and introduced the Marglobe in 1925. It was more disease resistant and had heavy foliage. It ripened slowly, and the fruit did not blister in hot sun, but it was an exterior ripener: appearing ripe on the outside, it could still be immature on the inside. Beginning in 1926 Campbell's plant pathologists teamed up with the New Jersey Agricultural Experiment Station at Rutgers University to cross interior ripeners with disease-resistant tomato varieties.[12]

THE RUTGERS TOMATO

The New Jersey Experiment Station had been established in 1880. At the time estimates were that the tomato industry generated an annual income in excess of one million dollars for the state. Due to the importance of tomato growing in New Jersey, the station worked continuously to improve tomato culture. In 1889 its first bulletin on tomatoes was authored by the director of the station, Edward B. Voorhees. Subsequently, station personnel conducted experiments on the effects of fertilizer, which demonstrated that nitrate of soda was an important fertilizer when coupled with phosphorus and potash. Tomato experimentation increased in the 1890s, when botanist Byron D. Halsted began experimenting with tomato varieties.[13] In 1896 the station experimented with growing tomatoes under glass. One experiment demonstrated that it was profitable to grow tomatoes under glass for commercial purposes. Maximum yield for greenhouse tomatoes resulted when plants were given two square feet of bench area. The results of these studies were published in the United States Department of Agriculture's Farmers' Bulletin no. 76, titled "Tomato Breeding" and authored by Voorhees.[14]

Halsted continued his breeding experiments. Nearly all commercial varieties of tomatoes were cultivated, and many were crossed. Two new varieties were developed: the Station Yellow and the Magerosa, a cross between the Magnus and the Ponderosa. The Giant tomato variety was also developed. In 1906 the station engaged in a major examination of over four hundred tomato varieties. Two years later Halsted focused on determining the best methods of growing tomatoes and disseminating that information to farmers. The station expanded its number of experiments in 1912 and examined the effects of fertilizer and tomato staking. In 1914 five thousand tomato plants were crossed.[15]

Lyman G. Schermerhorn was appointed to the New Jersey Experiment Station in 1914 and began a systematic investigation of tomatoes. Most of Schermerhorn's time was spent teaching, but he assigned senior students to conduct technical studies. More tomato varieties were crossed. One cross between the Marglobe and the J.T.D. seemed particularly promising.[16] Building on a cross initially made by Campbell's research staff, Schermerhorn cultivated literally thousands of plants resulting from this initial cross. He found four that showed potential by 1928. These were field-tested on farms located in New Jersey tomato growing regions. The result was a plant with the required canning characteristics of the J.T.D. and the disease resistance of the Marglobe. Schermerhorn called his new variety the Rutgers tomato.[17]

The Rutgers was a rather large plant with thick stems and large leaflets that protected the fruit from the hot sun. Its fruit was scarlet red in color, medium large in size, and slightly flattened in shape. It ripened from the inside out. Its juice was low in acidity and fine flavored. It was ideal for canning.[18] Schermerhorn officially introduced it in 1934. A press release issued by the New Jersey Experiment Station on September 19 of that year modestly asserted that the Rutgers tomato would yield "growers more money while providing consumers with a product of unexcelled quality." It yielded on average fourteen tons to the acre.[19] In 1935 nine thousand acres were planted, mainly for canning purposes. The Campbell Soup Company, "an ardent cooperator in selecting the strain," planted thirty acres in Cinnaminson to supply seed to its two thousand tomato growers. It also sent seed to Georgia to produce fifty million more plants.[20]

The Rutgers tomato was widely disseminated throughout the United States and the world. Canadian growers found that it could withstand drought better than any other variety then available. From Australia and South Africa came other good reports. In Hawaii it bore fruit at an altitude of three thousand feet on a mountainside. A Frenchman doubled his yield using the Rutgers tomato. One farmer claimed to have produced eighteen tons per acre. From 1935 to 1951, over one million pounds of certified Rutgers tomato seeds were distributed. As tomato seed often counts from seven thousand to fifteen thousand to the ounce, this was a lot of seeds by any standard. By 1952, the Rutgers tomato represented 72 percent of all commercially grown tomatoes in the United States and was widely cultivated in other countries.[21] Within two years, however, the Rutgers tomato was superseded in California by varieties that could be mechanically harvested. Although the Rutgers tomato remained an important variety in New Jersey, its days were numbered as California researchers assumed the leadership in the development of new varieties.

SOUP TOMATOES

In the early years, Campbell provided seeds to its contract growers. When Hall began growing seedlings under glass on thirty acres at Dorrance's home in Cinnaminson, plants were given to contractors.[22] They yielded twenty million plants. As Campbell's tomato canning operations expanded, even this was not enough. Based on an idea developed by R. Vincent Crine of Monmouth County, New Jersey, Campbell began growing open-air tomato plants in South Carolina in 1918. Crine moved the operations to Georgia in 1919. When the growing season commenced in New Jersey, the Georgia plants were shipped northward. Simultaneously, the hothouse and open-air plants were then distributed to Campbell's contract farmers. To maintain the highest quality, Campbell's field agents watched over the planting and growing operations of its contract farmers.[23]

The company was particularly interested in developing and maintaining good relationships with its growers. It convened meetings of growers and awarded prizes to those who did exceptionally well during the year. With the encouragement of Harry Hall, the company created a Ten-Ton Club in 1934 for its growers. Admission to the club was limited to those who grew at least five acres of tomatoes and generated ten tons of tomatoes per acre. During the first year only twelve growers qualified, but by 1951 membership had increased to 767. The club sponsored an annual meeting and a newsletter, which encouraged farmers to use the best fertilizers, crop rotation, spraying, and other up-to-date cultivation techniques.[24]

In addition to breeding new tomato varieties, the Campbell's agricultural research program was also interested in ways to protect the tomato crop from devastating diseases such as fusarium wilt. A new protective program resulted in higher-quality fruit for delivery to the canneries during the 1950s. During the next decade, the Campbell Soup Company released several new varieties, including Campbell 17 (C17-1965). This strain increased yield through better resistance to both fusarium and verticillium wilts.[25] Other supertomato varieties were under development at the University of California at Davis. In 1961 the university released VF145, which quickly became the dominant variety among California tomato harvesters.

Federal and state agricultural experiment stations expanded their tomato variety development work, releasing a series of multiple disease-resistant varieties for use or for breeding programs, particularly from work in Florida. Work also progressed on developing varieties resistant to insects. When the Tomato Genetics Cooperative was formed, it was initially based at the University of California at Davis. Its purpose was to exchange information and seed stocks used in genetic study.[26]

THE REVOLUTIONARY EFFECTS

The revolutionary changes noted above greatly influenced tomato growing and processing in America. Perhaps the most dramatic changes transpired in California. Tomatoes were grown in California well before its acquisition by the United States. Franciscan priests, who established most of the missions in the Spanish-controlled parts of what is now the state of California, grew tomatoes in their gardens. Father Fermín Francisco de Lasuén, stationed at Mission San Diego, received a donation of tomatoes from Mission San Gabriel near Los Angeles in 1777.[27] When Mexico revolted against Spain in 1813, California became a part of the Republic of Mexico. Its residents grew tomatoes throughout the early nineteenth century, and tomatoes constituted an important part of their cookery. The United States conquered California during the Mexican War. Shortly after the conquest, gold was discovered. The price of food rocketed sky high as most people preferred to prospect for gold rather than farm the land. For tomatoes, one estimate was that one and a half acres generated the equivalent of eighteen thousand dollars in yearly income.[28]

By 1851 California produced large tomatoes, if not large tomato crops. San Francisco's *Daily Alta California* proclaimed that "if anybody doubts the capability of California to produce vegetables of every description, and the largest possible size, let such an one take a walk down the Central wharf, any morning, and witness the display which is daily made there." Among the finds were "big luscious tomatoes, larger than the world ever before produced." The writer concluded correctly: "Blest, as we are, with a soil of such surprising richness, there need be no fear of California becoming uninhabited, even though the gold of the Sierra should be exhausted to-morrow."[29]

The *California Farmer* was launched in 1854 and regularly featured tomato recipes and articles on tomato culture. The California State Agricultural Society offered premiums for tomatoes by 1856. When the *California Culturist* was launched in 1858, it likewise spotlighted tomato culture and cookery.[30]

Tomato canning quickly took off. The discovery of gold in California gave the first great impetus to the packing of canned goods in the eastern states, and for years California was one of the largest and most profitable markets for eastern packers. But canning foods in the East and shipping them to California before transcontinental railroads or the Panama Canal meant that the goods first had to be shipped to Panama, offloaded, carried over the isthmus, and then reloaded on ships in the Pacific for transport to California. The only other alternative was the arduous trip around the tip of South America.

The high cost of shipping canned goods from the East encouraged the development of the canning industry in California, and many firms were established. The first known cannery in California was set up by Dan R. Provost, who arrived in California in 1854 to establish an outlet for the goods produced in the East by his brother's firm, Wells and Provost. Provost quickly began production in California. P. D. Code became the managing director of the office in 1856. In 1867 Code established P. D. Code & Company and began bottling tomato ketchup. Josiah Lusk established Lusk Canning in Oakland and sent goods to the East Coast via sailing ship around South America. Boston-born Francis Cutting purchased a half interest in a small vinegar works in Gold Rush City in 1858. He started canning tomatoes in 1860. By 1864 Cutting & Company packed tomatoes in two-pound, two-and-a-half-pound, and five-pound cans. By 1868 they packed five thousand cases of tomatoes annually. In 1898 the renamed Cutting Fruit Packing Company joined eighteen other packers representing about half of the other canners in the state to form the California Fruit Canners Association. This association subsequently changed its name to California Packers Association (CalPack) and marketed Del Monte Brand products.[31] In 1890 Joseph and William Hunt founded the Hunt Brothers' Fruit Packing Company in Santa Rosa. In 1896 they moved their operation to Hayward. The Hunt Brothers produced many tomato products but did not join the California Fruit Canners Association.[32]

On July 12, 1869, a momentous event occurred. Fresh tomatoes grown in California were shipped by railroad to a Charles Drake, living in New York. As the editor of the *American Agriculturist* dryly commented, the tomatoes "were picked too late for so long a journey, and were a little over-ripe."[33] Two years later, the editor of the *San Jose Patriot* foretold that "a fortune can be made in growing tomatoes in the San Jose Valley," as they matured four to six weeks before those in the Sacramento Valley.[34] In the same year, the *California Horticultural and Floral Magazine* announced that tomatoes grown in California were better than those grown in the eastern United States. Another article stated that tomatoes could be grown in California eight months out of the year.[35] The shipping of fresh tomatoes to the East Coast did not begin in earnest until 1900. By 1916 approximately twelve hundred train-car loads of tomatoes were shipped east annually.[36] The expansion of the nation's transportation system reduced the cost of transporting goods from California farms and factories to eastern markets and improved the food distribution network, making commercial canned goods and fresh produce available to nearly every town in America.

The transcontinental railroads also greatly spurred on the development of tomato canning in California. By 1875 California was a net exporter of canned goods. The *San Francisco Journal of Commerce* reported that the city had five firms engaged in canning and that sixty thousand cases of tomatoes were produced during the year. There were also canneries in San Jose and Oakland.[37] California slowly increased this level of tomato production during the following years.[38] By 1881 California's canned goods were dispatched not only to New York and Chicago but also to Great Britain, China, Japan, the Sandwich Islands (Hawaii), Mexico, Central America, South America, Siberia, and Australia.[39]

Farms sprang up in southern California as well. Tomatoes were grown commercially in Los Angeles by 1871. A ten-acre tomato farm located not far from Cahuenga Pass (today in the middle of Los Angeles) generated about two hundred dollars per acre growing tomatoes. According to several writers, southern California produced tomato crops right through the winter months. Fresh winter tomatoes were sent by stagecoach to San Francisco via Wells Fargo Company.[40] Further south in San Diego, tomato plants grew to gigantic proportions. Kate Sanborn claimed to have seen "a tomato vine only eight months old, which was nineteen feet high and twenty-five feet wide, and loaded full of fruit in January. A man picking the tomatoes on a stepladder added to the effect." This may not have been a completely accurate observation, for Sanborn, presumably tongue in cheek, titled her book *A Truthful Woman in Southern California*.[41]

Whatever the size of the plants, the amount of tomato fruit grown and canned in California escalated. In 1889 California canned a quarter of a million cases of tomatoes. In 1897 Edward Wickson proclaimed that the tomato was "one of the most popular, prolific, and profitable of California vegetables." Wickson reported that it was "grown everywhere during the local occurrences of the frost-free period, and in our thermal situations the fruit can be gathered all the year." Despite such lavish statements, California's love affair with the tomato had just begun. By 1907 the pack was close to one and a half million cases.[42]

Into the first decade of the twentieth century, wagons were the major source of conveyance for the tomatoes. From 1908 trucks increasingly were used to transport tomatoes from fields to factories. Trucks hauled larger amounts of tomatoes over greater distances in shorter periods of time, thus reducing spoilage and encouraging concentration in production. Large factories proved more economical and more efficient, and small factories were unable to compete.

California is uniquely suited for raising tomatoes. It has the right climate, an adequate access to water, and the farms large enough to make it economically attractive to invest in the major equipment associated with tomato farming. Technology also played a major role in the sensational increase of processing tomatoes in California. Before World War II, California produced 20 percent of the nation's tomatoes. The invention of the mechanical harvester eliminated the need for labor-intensive hand picking of the tomatoes. The mechanical harvester caught on in California during the late 1940s, but its real effects were not felt until the 1950s. By 1953 California growers cultivated 83,000 acres and produced 50 percent of all tomatoes in the United States. This had reached 130,000 acres by 1960.

It soon became evident that the harvester reduced the cost of processing tomatoes by two thirds. Between 1964 and 1968, the use of tomato harvesters increased from 3 percent to 95 percent. By 1970 California produced 70 percent of all processing tomatoes grown in America. But the growth was not over. By 1980 over 210,000 acres of tomatoes were grown in California, and ten years later this had increased to 320,000 acres. During this same time, yields greatly increased. In 1967 California growers harvested between seventeen and twenty tons of tomatoes per acre. By the 1990s they averaged thirty-four tons per acre. Hence, total production of tomatoes for processing increased from 2,250,000 tons in 1960 to over 9 million tons in 1990.[43] Today, 90 percent of all American-processed tomatoes are grown in California.

Revolutions often have explosive and unexpected results. This is also true of tomato revolutions—literally and figuratively. On November 4, 1991, in Camden, New Jersey, five hundred people watched as the old buildings that had housed Campbell's soup production were torn down. In many ways their demolition was anticlimatic. Campbell had stopped making soup in Camden in 1980, and all operations ceased ten years later. The demolition of the buildings was intended to make way for a major new construction project, which the city of Camden desperately needed. Not everyone was happy with their demise. Those who worked at the plant wistfully reminisced about the good old days. Others were keen on preserving as historical landmarks the gigantic Campbell's soup cans that adorned the top of the buildings. Their efforts failed: the soup cans were destroyed along with the rest of the building.[44] But the story did not end here. Campbell transferred the property to the city of Camden. Currently on the site is one of the finest Children's Gardens in the nation. Needless to say, vegetables including tomatoes are grown in the garden.

Tomato growing in South Jersey continued to flourish during the early twentieth century. In 1919 the state counted 37,000 acres of tomatoes under cultivation. Within a decade it had reached a peak of 42,000 acres. During the Depression, tomato cultivation slipped back to 1919 levels, but increased needs during World War II gave the tomato industry a new lease on life. By 1945, 55,000 acres of tomatoes were back under cultivation. Tomato cultivation then generated 10 percent of all agricultural revenues in New Jersey. But this was the tomato industry's apex. Total tomato acreage declined during the 1950s and 1960s. In 1955 total tomato cultivation acreage had dropped to 25,000 acres. Ten years later, it dipped to 19,000. The tomato breeding efforts that revolutionized the tomato industry in California also had an impact in New Jersey. Crop yields had increased from ten tons per acre to over twenty tons by 1962. But the real problem was cost. New Jersey tomatoes were still picked by hand, which cost three times more than harvesting mechanically. Climatic conditions in New Jersey were not as ideal as in California; fall rains lowered the quality of the tomato crop and made it impossible for mechanical harvesters to operate efficiently. In addition, land prices in New Jersey escalated, increasing the cost of tomatoes, and many growers sold out to land developers.[45] When processors moved their operations to California, the tomato industry in New Jersey virtually disappeared. By the 1990s almost all tomato growing in New Jersey was concentrated on producing fresh tomatoes for consumption in major cities, such as Philadelphia and New York.[46]

The Campbell Soup Company retains its world headquarters in Camden and remains the largest tomato soup manufacturer in the world, but its main tomato processing operations are now located in California. Seventy to 80 percent of processing tomatoes used by the Campbell Soup Company come from sixty contract growers who reside within a

150-mile radius of the tomato processing plants in Stockton and Dixon. Growers use about fourteen tomato varieties supplied by the company; which one varies based on soil types, length of time each variety takes to mature and other factors. Campbell's agricultural representatives monitor the progress of the tomato crops in the fields of their growers, specifically looking for quality and projecting the time to maturity. This information is fed into the Tomato Information Processing System (TIPS), which processes the raw data. From TIPS, projections are made predicting when tomatoes will be available for processing, how many trucks will be required to collect the tomatoes, how long it will take the trucks to arrive at the processing plant, and what the quality of the tomatoes will be when they arrive.[47]

At the proper time, mechanical harvesters scoop up the tomato plants and shake the tomatoes off the vines. Electronic eyes disgorge green tomatoes, while the ripe tomatoes are transported via a conveyor belt to a truck under contract with Campbell. The processing plants receive tomatoes from the Fourth of July to about the first of October, depending on the weather. The first stop for the tomatoes is the grading stations operated by the Process Tomato Advisory Board, a division of the California Department of Food and Agriculture. From the grading station, the tomatoes are delivered to the processing plant, which handles about 350 truckloads or about nine thousand tons of tomatoes per day. During peak season the plant operates around the clock.

From the trucks, the tomatoes are poured into a freshwater bath and are lifted by water flumes into the processing plant. All tomatoes are examined by professionally trained inspectors. Tomatoes that do not meet standards are removed by hand; the remaining tomatoes go into a chopper. The seeds and stems are then extracted. At this point the pulp is about 6 percent solids. Initially, most producers removed water from tomatoes by boiling. During the 1950s evaporators originally developed for the dairy industry were adapted for use in tomato processing. The evaporators rapidly remove water and concentrate the pulp to 42 percent solids. Some concentrate is frozen via flash coolers, which remove water and heat, as the paste falls through the machine. The chilled concentrate is then stored in fifty-five-gallon drums and used when needed. Other concentrate is pumped into aseptic bags, which exclude outside air. Framed in collapsible wooden boxes, the bags are loaded on trucks and shipped to the Campbell factory in Sacramento, or shipped in railroad cars to the factory in Napoleon, Ohio. In a strange twist, tomatoes are no longer grown in Ohio for the Campbell Soup Company; so efficient are the operations in California that it is more economical to grow and process tomatoes in Stockton and Dixon and pay the cost of shipping the concentrate across the country.

At the soup factory, the drums and bags of tomato paste are reconstituted. Spices and other ingredients are quickly blended in, and the soup is rapidly heated. Cans are immediately filled, sealed, and reheated for approximately fifteen minutes. Then the cans are rapidly cooled, labeled, and cased for shipment. As tomato soup is now made year round, the cases are shipped as fast as possible to retailers.

Although tomatoes may be harvested up to 150 miles away, the total time from the harvesting to the departure of the paste from the California factory rarely exceeds seven hours. Once tomatoes arrive at the plant, they are usually processed within an hour and a half. From each ton of fresh tomatoes received into the processing plant, twelve thousand cans of tomato soup are produced at the end of the process.

Professionals can determine the difference between soup made with fresh tomatoes and soup made with concentrate. However, it is almost impossible for consumers to tell the difference. This "just in time" process is a more efficient system, and consequently costs are lower. It also produces a better-quality product for the consumer. In the past, all the tomato soup was hectically made during the months of August and September, and the quality varied depending on the tomatoes processed and weather conditions. Today tomato soup produced in Napoleon in December is the same as the product manufactured in Sacramento in May. In the past, the tomato soup manufactured in August sat around in a warehouse, sometimes for almost a year, until distributed to retailers. Now, as tomato soup is made when needed, the end product is fresher when it arrives in the grocery store.

THE SUPERTOMATO

As significant as the tomato revolutions of the past have been, they may pale when compared to future changes. The driving force on California's road to tomato stardom has been the University of California at Davis and its emphasis on basic research into the tomato. The efforts originally centered on the Tomato Genetics Cooperative (TGC), which was founded in the fall of 1950 by two graduate students, Don Barton and Allen Burdick, and Charles Rick, the University of California at Davis's geneticist. The purpose of the TGC was to "exchange information on tomato genetic research, stimulate linkage studies and preserve and distribute germplasm." The first report of the TGC was issued in 1951. Rick edited the *TGC Report* until 1981. Under the direction of a Coordinating Committee, one of the TGC's more important activities was the creation of a Gene List Committee. It was assigned the task of compiling and publishing lists of known tomato genes and revising symbols when appropriate to conform to nomenclature rules. During the following decades, hundreds of genes were identified and categorized. Along with the identification of the genes came the listing of sources of seeds for each gene. Identification of the location of the seeds became cumbersome as the number of known genes expanded. Rick established the Tomato Genetics Stock Center in 1976.[48]

To prevent duplication of research, different chromosomes were assigned to different investigators. The first linkage map was published by the TGC in 1955. It identified forty-seven genes on eleven chromosomes. In 1977 a revised map included 288 genes, and each of the twelve chromosomes was mapped for marker genes.[49]

Tomato genes have been mapped extensively during the past thirty years. Each of the tomato's twelve chromosomes can be easily recognized during certain stages of its reproduction. One noteworthy genetic trait of the tomato is that specific functions are controlled by only one gene. This combination of characteristics has made the tomato a relatively easy organism in which to locate exact chromosomal positions, which greatly aids bioengineering techniques. Bioengineers maintain that through selective breeding and naturally occurring mutations they have produced genetically engineered tomatoes with some of the qualities that breeders have sought for years.

Beginning in the 1950s, botanists induced genetic mutations with X rays and chemicals. These mutants were mainly of interest to researchers. The research, however, encouraged further investigations into the chromosomal structure of the tomato, making more sophisticated alterations possible. During the past few years this research has become productive. In a project funded by Campbell's Soup Company, Calgene, Inc. in Davis, California, produced the first genetically engineered tomato, called MacGregor's. Calgene claims that MacGregor's is slow ripening, "terrific tasting," and transportable over great distances without loss of quality. The variety has received a patent, and the U.S. Food and Drug Administration (FDA) has permitted it to be sold to the public. Other genetically engineered tomatoes are under development. Another bioengineering firm, DNA Plant Technology of Cinnaminson, New Jersey, has "synthesized genetic material from fish that showed promise in protecting tomatoes from freezing." Other companies with strong biotech programs, such as Petroseed, Monsanto, Pioneer, and Dupont, are not far behind.[50]

Calgene conducted extensive testing on another bioengineered variety called Flavr Savr and voluntarily presented its genetically altered tomato to the FDA for approval. Hearings were held, and David A. Kessler, the commissioner of food and drugs, said, "I heard no dissent today on the safety evaluation of the Flavr Savr tomato, and the committee thought that the evaluation of the Flavr Savr was exceptionally thorough." The genetically engineered tomato sailed through the FDA's Food Advisory Committee. Calgene produced a substantial amount of scientific data to show that it was safe. Thomas C. Churchwell, president of the subsidiary Calgene Fresh Inc., stated that the rapid softening of ripe tomatoes was caused by an enzyme called Polygalacturonase, or PG. "PG breaks down the pectin in the tomato's cell walls so it will decay faster and allow the fruit's seed to spread on the ground. Calgene spliced into the tomato's genes an extra one to cancel out the effect of the PG enzyme by about 99 percent. Thus, the tomato remains firmer in its last week and can be left on the vine to ripen for an extra two or three days, until it begins to flush red. It can then be picked and allowed to ripen further en route without gassing to produce redness. After a week or so of extra firmness, the Calgene tomato softens and decays like other tomatoes." Critics complained that Calgene scientists had "inserted a second gene to mark the successful insertion of the one to cancel the effect of PG. This marker gene, called Kan-r, confers resistance to the antibiotic kanamycin." Critics believed that the bioengineered tomato could pass the Kan-r gene to people, "negating kanamycin's effectiveness as a prescription drug to stop an infection." Calgene presented data demonstrating that the Kan-r gene was quickly denatured in the stomach and ineffective in neutralizing the antibiotic.[51]

Others, including the futurist Jeremy Rifkin, have strongly opposed genetic engineering of the tomato as presenting an unacceptable risk for humans. Some grocery stores have refused to sell genetically engineered tomatoes, while others have agreed to identify them as genetically engineered. Restaurants have announced that they will boycott the new "mutant" tomatoes. A major concern is that there is no requirement that genetically altered foods be labeled as such. Many critics believe consumers have the right to know which products have been altered and which are natural. On May 18, 1994, the FDA approved Flavr Savr. Since then, genetically altered tomatoes have been sold in grocery stores.

In other countries, those opposed to genetic engineering have taken more direct steps to oppose genetically altered crops. In France, two hundred protesters swooped down on a test site belonging to U.S.-based Monsanto Company and ripped up a showcase of bioengineered crops. In the United Kingdom, schools banned genetically modified foods from their cafeterias because of "health concerns."

While current controversies have focused on the genetic engineering of tomatoes for the fresh market, it is extremely likely that major corporations are also using the technique to improve the taste, color, processability, and disease resistance of tomatoes for soupmaking.

■ CHAPTER 7 ■

An American Culinary Icon

In 1962 the Ferus Gallery in Los Angeles displayed Andy Warhol's thirty-two paintings of Campbell's soup cans, each representing a different variety. The paintings were displayed in a row on horizontal white shelves wrapping around the gallery's wall, suggesting the appearance of soup cans on a supermarket shelf. At the same time, their frames insisted on their identity as works of art.[1] Neither art critics nor the viewing public knew quite how to respond to the show. According to the gallery owner, Irving Blum, visitors were extremely mystified. Local California artists "were provoked by these paintings but they tended to shrug, not really condemn. There was a lot of amusement. A gallery dealer up the road bought dozens of Campbell's soup cans, put them in the window and said, 'Buy them cheaper here—60¢ for three cans.' That was publicized. So there was a lot of hilarity concerning them. *Not* a great deal of serious interest." In another interview Blum reported that the reaction "was very peculiar. There was a lot of amusement. People felt that they were somehow slightly ridiculous."[2]

Born Andy Warhola in Pittsburgh in 1928, Warhol dropped the *a* at the end of his name while still in college. In 1949 he moved to New York, where he became a successful commercial artist designing advertisements for women's magazines. Despite his financial success, Warhol wanted to do "serious" work that would be presented in a major exhibition. In the summer of 1960, he conducted two experiments. One was the painting of two six-foot high Coke bottles. The other was a series of cartoon characters. Roy Lichtenstein, then living in Highland Park, New Jersey, picked up the cartoon idea and promptly cashed in on it. When Lichtenstein's works were exhibited and received national acclaim, Warhol went in search of new inspiration. At the end of 1961 Warhol asked advice from Muriel Latow, an interior designer who owned a struggling art gallery. Latow asked Warhol what he liked most in the world. Warhol replied,

"Money." Then, she responded, "you should paint pictures of money." She further explained: "You should paint something that everybody sees every day, that everybody recognizes, like a can of soup."[3]

Latow demanded and received fifty dollars for her idea. Her advice had struck memory's chord in Warhol. The following day he asked his mother to purchase every Campbell's soup variety available in the local store. Warhol painted "portraits" of each can as precisely as possible against a white background. The one major exception was that his cans did not possess the gold medallion won by the Joseph Campbell Company in Paris. This medal featured two reclining allegorical figures, one male and one female, and the date of the award—1900. During the first half of 1962, Warhol continued working on the soup cans.[4]

Irving Blum visited Warhol in May 1962. He spent quite a while studying the soup can paintings and decided to organize an exhibit of all thirty-two. When he presented the idea, Warhol responded, "Oh, I would adore it." Each painting was to sell for one hundred dollars, and Warhol was to receive half. At the time Warhol was earning ten times this amount per illustration for his commercial work. When the paintings arrived in Los Angeles, Blum mailed out a postcard announcement with the image of a tomato soup can. The exhibit opened on July 9, 1962, but there was no formal opening. Warhol never did visit the exhibition.[5] Blum sold half a dozen paintings, including one to Dennis Hopper. Then Blum decided that he wanted to keep the entire group, so he reacquired those he had sold. According to Blum, the soup cans "were extremely provocative." He called Warhol, who was thrilled "because they were conceived as a series. If you could keep them together it would make me very happy." For the complete group, Blum paid Warhol one thousand dollars in installments over the next year.[6]

Even before the Campbell's soup cans were displayed at the Ferus Gallery, they had achieved notoriety. Warhol was interviewed by *Time* magazine. According to the article, which appeared on May 11, 1962, Warhol was "currently occupied with a series of 'portraits' of Campbell's Soup cans in living color." Warhol stated that he was painting "the things he always thought were beautiful, things you use every day and never think about." He continued: "I just do it because I like it."[7] On another occasion he expanded his comment: "I used to drink it. I used to have the same lunch every day, for twenty years, I guess, the same thing over and over again."[8] He stated that "many an afternoon at lunchtime Mom would open a can of Campbell's tomato soup for me. . . . I love it to this day."[9] While Warhol painted many different Campbell's soups, the tomato soup can became his most famous image.

The paintings were assaulted in the press when the exhibit opened. As word spread, people took strong positions pro and con.[10] The editor of

Artforum, John Coplans, visited the exhibition and was impressed. On the other hand, "What could be more absurd," asked Ralph Rugoff, "than an assembly-line commodity that comes to stand for home-cooking and motherly love, only rendered virgin and immaculate?" When asked by a New York dealer what the difference was between his soup paintings and Rothko's, Warhol replied: "Mr. Rothko signs his name on the back of his product and Campbell's does it on the front." But, hinted Warhol, "both were signature products."[11]

Warhol hand-painted Campbell's soup cans later in 1962, including the famous torn cans. He employed silkscreens in 1965. Twenty years later David Bourdon praised the eloquent and tragic torn soup cans as "imitations of mortality."[12] The soup cans exist in a striking range of size, varying from a monumental six feet to portable pencil sketches. They appeared in isolation and crowded in grids. They were of various colors. Unlike the first set, subsequent renditions sometimes accurately portrayed the medallion. The original omission and the subsequent inclusion led art critic Steve Jones to conclude: "The removal of the two human figures and the historicizing date produce two distinct advantages. The differentiation of the commercial product from the iconic image to become a point around which the conservative function of memory and the transformative forces of historical process can coalesce. Additionally, the evacuated medallion becomes a locus which the viewer can inhabit as a site of the self."[13]

When his soup-can paintings were ridiculed, Warhol professed not to care. As one observer reported, "It was all good advertising. It gave him free publicity."[14] In January 1963 Donald Judd asked: "Why does Andy Warhol paint Campbell Soup cans?" Judd's answer was "Why not?" To Judd, the soup cans were "a cause for both blame and excessive praise."[15] In 1965 Marianne Hancock of *Arts Magazine* reported that Warhol's "icy portraits of tomato-soup cans rouse strong and sometimes hostile emotions."[16] Warhol generated phenomenal public support. Many fans sent him letters on labels from soup cans.[17]

Warhol's soup cans remained controversial. The critics heatedly debated their artistic merits. Some asked, "Is this art?" Ivan Karp addressed this issue when interviewed at the Society of Illustrators meeting in 1965, asserting that Warhol had taken the Campbell's tomato soup can "out of its context in a supermarket, by blowing it up, putting it on canvas, and adding something ineffable of his own." Warhol "transfigured" the can "and raised it to the level of spiritual importance. By his act of transmutation he created beauty and meaning." *Esquire* magazine, always ready to "deconstruct" celebrities, proclaimed Warhol's soup cans to be the symbol of the whole pop art movement. One *Esquire* cover showed Warhol drowning in a can of tomato soup with the caption: "The final decline and total collapse of the American avant-garde."[18] Yet the article associated with the cover concluded that anyone who could "parlay a soup into personal stardom" was "hot stuff." But whatever the controversies, as Henry Geldzahler wrote, "Warhol captured the imagination of the media and the public as had no other artist of his generation."[19]

In Warhol's exhibition at the Whitney Museum, his early work was excluded. The exhibition began with his Campbell's soup cans. According to *Art* critic Lawrence Alloway, the soup cans gave "the *appearance* of anonymity and the fact of repetition make them characteristic of Warhol's main line as a painter. The silk screen is central to his art and it is used like this: an image is mechanically transferred to a screen which can then be used repeatedly. The amount of paint and the pressure with which it is applied affects the impression that the screen leaves on the canvas."[20]

In reviewing *The Philosophy of Andy Warhol* in the *New York Times Book Review,* Barbara Goldsmith reported that Warhol's visual images of Campbell's soup cans "shoved art out into 'the real world'—a world of commercialism and mediocrity."[21] The Campbell's soup cans were an icon of mass production. It was the one item likely to be found in the cupboards of every kitchen. As Steven Kurtz wrote: "One image of a Campbell's soup can must be equivalent of the image of all other Campbell's soup cans. Image consistency is as important as product consistency. Through this strategy of desire, products may be endlessly replaced without any feeling of loss or deprivation. Campbell's soup flows in an endless stream that has no beginning or end. Unlike a fine wine, which is marked by finitude since it is irreplaceable when the stock is depleted, Campbell's soup flows in the realm of infinity—the soup I ate, and Andy ate as a child, is the soup I eat today. Andy's experience and my experience are equal in the realm of simulation. There is no original."[22] The cans evoke memories, reported art critic Steve Jones, "involving both the history of a culture and that of each individual within that culture."[23]

The Campbell Soup Company at first did not know quite how to respond to the images of its flagship product exhibited as works of art. The company commissioned Warhol to paint a soup can for its board chairman in 1964. But when the Institute of Contemporary Art in Boston exhibited Warhol's work in 1966 and sold a shopping bag with an image of Warhol's Campbell's tomato soup can, lawyers for the Campbell Soup Company wrote a letter to the producers of the bag, expressing concern about the company's trademark appearing on "any commercial products without our specific consent" and demanding that production of the bags be stopped immediately.[24]

The shopping-bag controversy soon blew over, but this did not end the company's connection with Warhol. Licensed under agreement with the company, manufacturers have reproduced Warhol images with his signature on T-shirts, sweat suits, coffee (or soup) mugs, pitchers, glasses, and Christmas tree ornaments. In 1968 Warhol himself designed a dome-shaped "Bread Loaf" lunch box, complete with a red-and-white thermos and a tomato soup can label on the outside.[25] For the hundredth anniversary of the Campbell's condensed soup, Steve Kaufman, purportedly a former assistant to Warhol, was commissioned to paint more Campbell's soup cans and jars. Two series of three screenprints composed of tomato, chicken noodle, and cream of mushroom soup cans on canvas "with hand embellishments" cost two thousand dollars.[26]

These were not the first images of tomato soup to appear on promotions. Besides advertising in magazines, on streetcars, and in newspapers, Campbell Soup Company has manufactured or licensed numerous products since the early twentieth century. Tomato soup can banks, can openers, money clips, ceramic utensil holders, burner covers, glasses, thermoses, magnets, pails, radios, popcorn tins, trash cans, mouse pads, posters, T-shirts, and watches are only a few of the estimated two thousand Campbell Soup Company collectibles. In 1995 the Soup Collector Club was founded by David and Micki Young of Wauconda, Illinois. The club held its first convention in Napoleon, Ohio. David Young, who has been collecting Campbell Soup collectibles for fifteen years, estimates that there are about a thousand other collectors across the country. The Youngs published the first Campbell Soup price and identification guide in 1998; it includes descriptions of about sixteen hundred items, many of which feature tomato soup. The club has about four hundred members, all of whom receive a newsletter with the latest information about the company's licensed products.[27]

A second Campbell's collectors' club originated in 1998. It held its first convention in Cherry Hill, New Jersey, not far from the Campbell Soup Company's world headquarters in Camden. Led by Patti Campbell of Ligonier, Pennsylvania, CC International, Ltd., is a not-for-profit organization composed of 235 members. It publishes a newsletter called the *Souper Gazette.*[28]

Nancy Bailey of Delmont, New Jersey, is reportedly holding one of the largest collections of Campbell's collectibles. Her collection of four thousand items has numerous tomato soup images, including ones on toy banks, trucks, ice buckets, a magnet with timer, a jacket, bicentennial dolls, a siren, a cigarette lighter, and sand pails.[29]

Thirty-three years after Warhol painted his first Campbell's soup can, the Campbell Soup Company sponsored a contest for the nation's youth,

"Art of Soup." The expressed intention was to inspire the next Andy Warhol. The company was shocked when a hundred thousand potential entrants called its special telephone line requesting entry information. Over ten thousand applied. Entrants were divided into amateur, professional, and youth status, and winners, announced in 1995, were chosen in six categories: Painting/drawing; sculpture; film/video; photography; crafts/wearables; and all other media. The eleven-year-old grand-prize winner, Matthew Balestrieri of San Juan Bautista, California, received ten thousand dollars, and three first-place winners received five thousand dollars each. Andy Warhol once said that everyone would have fifteen minutes of fame; all winners of this contest received more than their allotment, as their works were exhibited at the Whitney Museum of American Art in New York and promoted around the country by the Campbell Soup Company. According to Andy Warhol's brother Paul Warhola, "Campbell's search for new pop artists is wonderful because it's important to encourage and recognize unknown artists. If Andy were alive today, he'd probably enter the contest for the fun of it!"[30]

Campbell's soup cans were culinary icons well before Andy Warhol painted them in 1962. Over the past one hundred years, the Campbell Soup Company carefully created brand and label recognition among consumers. It was extremely successful: most Americans easily associated the red-and-white label with the Campbell Soup Company. But prior to Warhol's work, commercial art was divorced from the rest of the art world. Warhol helped make the connection between the two and led viewers to see everyday reality in new ways. Warhol's paintings and silkscreens enshrined Campbell soup cans as societal icons transcending time and commercial marketing techniques.

LADLE WARS

Soup has also made it into the theater and even onto prime-time television. Perhaps the first tomato soup to appear in the theater was a reflection of a real one offered by the self-service automat Horn & Hardart. The first coin-operated restaurant was opened by Joseph B. Horn and Frank Hardart in Philadelphia in 1902. The idea spread throughout Philadelphia and in 1912 came to New York. By the 1920s Horn & Hardart had become famous throughout America. During the Depression it was known particularly for its "tomato soup" composed of two free items—hot water and ketchup—favored by struggling actors and others short on funds. They purchased a tea bag and, after drinking the tea, filled the mug with ketchup and hot water. Ketchup soup was featured in Irving Berlin's *Face the Music,* a popular 1932 musical that included the song "Let's Have Another Cup of Coffee." In 1971 *An Afternoon at Hays Bickford* incorporated a scene with two old people consuming ketchup soup at Horn & Hardart.

But these allusions paled by comparison to the onslaught of soup visibility that began in the fall of 1995. America's most popular sitcom series, *Seinfeld,* aired episode 115, which introduced the "Soup Nazi." It portrayed a temperamental man who runs a soup restaurant and insists that all sales take less than seven seconds. The Soup Nazi's establishment usually has a long line of customers waiting to place their orders, and procedures have been established to keep the line moving: customers are supposed to order in a loud clear voice, pay for the order, move to the extreme left, and receive the order, all without unnecessary discussion and comment. If customers follow proper procedure, everything proceeds smoothly. Failure to follow these simple instructions could mean "No soup for you." At the episode's climax, Jerry Seinfeld's friend Elaine is thrown out of line for violating the rules. Jerry decides to remain at the soup stand. Elaine responds: "So, essentially, you chose soup over a woman?" Jerry dryly retorts, "It was a bisque." While the episode was fictional, the Soup Nazi was inspired by Al Yeganeh, who runs a takeout operation called the Soup Kitchen International on West Fifty-fifth Street in New York. It is but one of many eating establishments now specializing in serving soup. Soon after the Seinfeld episode aired, Chris Ramani opened the Soup Nutsy, a knockoff of the Soup Kitchen International, on East Forty-sixth Street. Its Manhattan Clam Chowder and its Tomato and Clam Broth were highly rated by the *New York Times* correspondent Eric Asimov.[31]

During the past few years several chains of soup restaurants have been launched. The first New England Soup Factory was opened in Brookline, a suburb of Boston, in 1995 by Marjorie Drucker and her husband, Philip Brophy, both graduates of the Johnson and Wales culinary program in Providence, Rhode Island. At first they sold about ten to fifteen gallons per day. When they opened their second in Newton, Massachusetts, they averaged 150 gallons per day. They have created about a hundred different soups, and these circulate through their outlets on a regular basis. Some soups are standards. Among the favorites is their tomato soup, a combination of tomatoes, rice, and basil.[32]

Another chain, Daily Soup, was the brainchild of Peter Siegel, Robert Spiegel, and Carla Ruben from Princeton, New Jersey. Their first store opened its doors in 1995. Since then, several additional locations have opened in New York City. The owners believe that "soup is a meal." Their soups are characteristically thick and hearty. Daily Soup has five hundred different types of soups, which regularly rotate through the establishments. The executive chef is Leslie Kaul, who received her formal training at the French Culinary Institute in Manhattan. According to a review of Daily Soup in the *New York Times*, a dozen soups are available daily. Some were novel, such as Malaysian Cabbage with Peanut; others were standard soups.[33]

The Daily Soup establishment's decor was described in the *New York Post* as "minimalist aluminum-and-white-tile." The reviewer continued, "They have simple black and white photos on the walls, print custom menus daily on paper bag colored stationery using a trendy font, offer a 'Happy Hour' special from 4:00 P.M. to closing, accept fax orders, and accept delivery requests via the Internet" at www.DailySoup.com. In addition Daily Soup gave "a piece of peasant bread, fruit and a cookie with every meal."[34]

Yet another chain, Hale & Hearty Soup, was launched by Douglas Boxer. It currently has eight outlets in New York.[35] In addition, street vendors ply the sidewalks of the city selling hot or cold soup, depending on the weather. Other establishments serves communities outside New York. In Seattle, the Sourdough Soup Company has gotten off the ground, while the Soupbox in Chicago is planning to go national with a soup-chain franchise. The Soup Kitchen Restaurant serves Kalamazoo, Michigan.

While soup kitchens in the tradition sense—places where the poor are fed—have survived, most now distribute a wide variety of foods. Tomatoes remain a popular ingredient in such kitchens, mainly because they are healthful and relatively inexpensive. In January 1999, a "Souper Week" was launched by some of New York's finest restaurants, who poured "their heart and soul" into preparing the "Best Soup in the World." In the tradition of Alexis Soyer, these restauranteurs sold their specialty soups at regular prices but contributed one dollar per dish to Citymeals-on-Wheels and Share Our Strength, nonprofit organizations that feed the ill and underprivileged.[36]

SOUPER BOWLS

In Gallup polls conducted between 1947 and 1973, respondents chose tomato soup as their first choice for the first course of their perfect meal, regardless of the expense.[37] The Campbell Soup Company estimated that it had produced its twenty billionth can of tomato soup in 1990. It declared tomato soup as the company's all-time best-selling variety. Today, tomato soup's popularity has waned. It is no longer the first choice of Gallup poll respondents. Neither is it Campbell's best-selling soup; it has been deposed by chicken noodle.

Yet tomatoes are ingredients in many of Campbell's soups, including Old Fashioned Tomato Rice, Low Sodium Tomato with Tomato Pieces, and Healthy Request Tomato. New products, including Campbell's

Ready-to-Serve Tomato Soup and microwaveable Campbell's Soup to Go, advertised as a "single-serving bowl for workplace and school lunches," have shown strong early growth.

Whatever their status at Campbell, soups with tomatoes are thriving. *Gourmet* magazine regularly publishes tomato soup recipes as well as soup recipes with tomatoes as an ingredient. During the past half century, these have included Iced Tomato Soup, Cream of Tomato, Consommé Brunoise aux Tomates, Cucumber and Tomato Soup, Cold Tomato Soup, Ripe Tomato Soup, Clear Soup, Paradise Soup, Chilled Tomato and Orange Soup, Fresh Tomato Purée for Fish Soup, and many more.[38]

Tomato soups from ethnic and immigrant groups have come to the fore. In the late nineteenth century, before it was generally accepted in Spain, gazpacho gained acceptance in Anglo-American high society. As one Spaniard observed, "With us, dogs but tolerably well bred would refuse to compromise their noses in such a mixture. It is the favourite dish of the Andalusians, and the prettiest women, without fear, swallow at evening great spoonfuls of this infernal soup." In 1896 Elizabeth Pennell reported that gazpacho was "most refreshing, an opinion which to us seems a trifle daring, and yet, extraordinary as it may be found at the first taste, you finish by accustomizing yourself to it, and even liking it."[39] Reports of its excellence had filtered back to the United States. When Alice B. Toklas expressed interest in acquiring a cookbook with gazpacho recipes, she was informed by a clerk in a Spanish bookstore that gazpachos were eaten only by peasants and American tourists. This response was one reason why she wrote her own cookbook, which included several gazpacho recipes.[40] Regional variations of gazpacho permeate Spain and other Spanish-speaking countries. Despite this diversity, the only gazpacho recipes that thrive in the American mainstream are those that are tomato based. As they are served ice cold, they may have been one of the first popular iced soups. Unlike in Toklas's time, recipes for gazpachos abound in cookery books and magazines today.[41]

After World War II the Mexican-American influence already evident in mainstream cookbooks became stronger, and immigrants from Mexico continue to bring to the United States their rich and diverse cookery traditions. Many of these traditions are now finding expression through the Mexican-style restaurants that are proliferating across America. Recipes for *albóndigas*, or meatball soup, abound in cookbooks and on restaurant menus.[42]

Likewise, Italian restaurants can be found in every city in America, and Italian soups are common in diners and coffee shops across the country. The first located Italian cookbook was not published in America until 1912, and few were published before World War II. Since the war Italian cookbooks have been regularly published, and most contain soups with tomatoes.[43] Many other ethnic and national culinary traditions encompassing soups with tomatoes have lately influenced American cookery, as well.

The total value of the commercial canned soup market in the United States increased from $2.3 billion in 1992 to $2.8 billion in 1996; Packaged Facts projects it will reach $3.6 billion by 2001.[44] Today, more than eighty American manufacturers produce soup in some form. However, only five produce more than 1 percent each of America's total soup output. The H. J. Heinz Company, for instance, produces 85 percent of all private label soups. Progresso Food Quality Products, still located in Vineland, New Jersey, is now a subsidiary of Grand Metropolitan PLC. Progresso makes a full line of ready-to-serve soups, including Tomato with Basil. The American soup market, however, is dominated by the Campbell Soup Company, which accounts for approximately 60 percent of all soup sales. It boasts four of the top five sellers in the canned soup market, one of which is tomato soup.

The Campbell Soup Company is the current victor in the souper bowl competition. It is the world's largest maker and marketer of soup, with fiscal 1998 sales of $6.7 billion. The company's soups are sold under the global "Campbell's" brand, "Swanson" broths in the United States, "Erasco" in Germany, and "Liebig" in France. Its products are sold in 120 countries. The company would not have achieved its dominant position without tomato soup. It is fitting to end with a quote from its current president, Dale F. Morrison: "My all-time favorite meal is a bowl of Campbell's Tomato Soup and a grilled cheese sandwich." Tomato soup "really is America's favorite food."

Historical Recipes

Tens of thousands of recipes for preserving or canning tomatoes and recipes for tomato soups or soups with tomatoes as an ingredient have been published in cookbooks, agricultural and horticultural journals, newspapers, almanacs, medical journals, magazines, and a host of other sources. In this section is a representative sample of those recipes. Some were selected because they were typical, others because they were unusual. Some recipes were experiments that did not survive, while others survive today. As a collection they reflect diversity and demonstrate the variety of sources in which tomato soup recipes appeared.

These historical recipes differ markedly from those that appear in modern cookbooks. Because of the state of cooking technology in the early nineteenth century, cooks were unable to control temperatures easily, and thus it was difficult to specify exact cooking times. Quantities often depended upon what was available. As tomatoes were not uniform, quantities depended too upon their size, shape, and consistency, as well as on such other factors as their acid content and their taste and color. When quantities were listed, they were frequently very large. Proportionally scaling back the quantities in these recipes will often result in finished products that taste similar to the originals.

Spelling and directions in these recipes have been left in their original form. There are a few measures and terms in them that are not commonly used today. For instance, a *gill* is a liquid measure equal to a quarter of a pint. A *peck* is equal to eight quarts. A *bushel* is equal to four pecks, or thirty-two quarts. A *tammy* is a cloth used for straining liquids. As ingredients have changed over the years, readers should be aware that it is almost impossible to recreate these historical recipes in modern settings.

RECIPES FOR TOMATO SOUP

(1770)

Take ripe Tomatas, peel them, and cut them in four and put them into a stew pan, strew over them a great quantity of Pepper and Salt; cover it up close and let it stand an Hour, then put it on the fire and let it stew quick till the liquor is intirely boild away; then take them up and put it into pint Potts, and when cold pour melted butter over them about an inch thick. They commonly take a whole day to stew. Each pot will make two Soups.

N.B. if you do them before the month of Oct they will not keep.

Source: Harriott Pinckney Horry Papers, Receipt Book, 28, South Carolina Historical Society no. 39-19; Richard Hooker, ed., *A Colonial Plantation Cookbook: The Receipt Book of Harriott Pinckney Horry, 1770.* (Columbia: University of South Carolina Press, 1984), 89.

Tomato Soup (1824)

Cut the stalks of two quarts of quite ripe tomatos; put them in a stew-pan with three onions sliced, one carrot, one head of celery, half a pint of good gravy; let it stew till tender, rub it through a tammy, and add three pints more good gravy, and season with salt and Cayenne pepper. Note.—A spoonful of beetroot-juice may be added if not red enough.

Source: John Conrade Cooke, *Cookery and Confectionary* (London: B. Bensley for W. Simpkin and R. Marshal, 1824), 11.

Tomato Soup—E.R. (1842)

Cut up an onion and fry it in a table-spoonful of oiled butter; when the onion is brown, take it out and replace it with a dozen tomatos, which must be stewed very gently until quite tender, pulp the tomatos through a sieve, and add them to the stock, which must be ready strained and thickened.

Source: Maria Eliza Rundell, *A New System of Domestic Cookery Formed upon Principles of Economy*, ed. Emma Roberts, 66th ed. (London: John Murray, 1842), 18.

Tomata Soup (1846)

Take a dozen unpealed tomatas, with a bit of clarified suet, or a little sweet oil, and a small Spanish onion; sprinkle with flour, and season with salt and cayenne pepper, and boil them in a little gravy or water; it must be stirred to prevent burning, then pass it through a sieve, and thin it with rich stock to the consistency of winter pea-soup; flavour it with lemon juice, according to taste, after it has been warmed up and ready for serving.

Source: [Judith Montefiore], *The Jewish Manual; or, Practical Information in Jewish and Modern Cookery with a Collection of Valuable Recipes and Hints Relating to the Toilette* (London: T. & W. Boone, 1846), 10–11.

Dry Tomato Soup (1846)

Brown a couple of onions in a little oil, about two table-spoonsful or more, according to the number of tomatoes; when hot, add about six tomatoes cut and peeled, season with cayenne pepper and salt, and let the whole simmer for a short time, then cut thin slices of bread, and put as much with the tomatos as will bring them to the consistency of a pudding; it must be well beaten up, stir in the yolks of two or three eggs, and two ounces of butter warmed; turn the whole into a deep dish and bake it very brown. Crumbs of bread should be strewed over the top, and a little warmed butter poured over.

Source: [Judith Montefiore], *The Jewish Manual; or, Practical Information in Jewish and Modern Cookery with a Collection of Valuable Recipes and Hints Relating to the Toilette* (London: T. & W. Boone, 1846), 97.

Tomato Soup (1880)

Put a gallon of water into the soup-pot, and add to it the bones and gristle left from a roast of mutton or any other roast. Boil until the liquid is reduced to two quarts; then strain, and set in a cool place. At the end of an hour skim off all the fat, and return the stock to the soup-pot, adding to it a can of tomatoes, four cloves, half a teaspoonful of pepper, a table-spoonful of salt, and a grain of cayenne. Put on to boil. In the mean time put four table-spoonfuls of butter into the frying-pan, and when it has become melted, add an onion, two slices of carrot, two of turnip, and four stalks of celery, all cut fine. Cook slowly for a quarter of an hour; then draw the vegetables to the side of the pan, and after pressing the butter from them, put them into the soup. Into the butter remaining in the frying-pan put four table-spoonfuls of flour, and stir until smooth and frothy; then add to the soup, being careful to scrape every particle of the mixture out of the pan. Let the soup simmer for half an hour. Taste, to ascertain whether there is enough seasoning. Strain, and serve with fried bread.

Two quarts of stock may be substituted for the bones and water.

Source: Maria Parloa, *Miss Parloa's Kitchen Companion* (Boston: Dana Estes, 1880), 141.

Tomato Soup (1884)

Boil one quart of ripe tomatoes in one quart of water ten minutes; then add one teaspoonful of salt, one teaspoonful of pepper, one teaspoonful of white sugar and a piece of butter the size of an egg; lastly add one quart of milk and let all simmer.

Source: Frances Willey, *The Model Cook Book* (Troy, New York: E. H. Lisk, 1884), 19.

Tomato Soup for the Sick (1886)

Take three quarts of tomatoes, canned or whole, and boil them until they are soft; then strain through a colander, afterwards through a flour-sieve, and reject the solid portions. Boil three pints of milk, thicken it with three tablespoonfuls of flour, then boil the liquid part of the tomatoes over again, and then stir this boiling liquid into the milk; put in a little butter; salt to taste.

Source: *Gardener's Monthly* 28 (July 1886):207.

Tomato Soup (1891)

This is a dish susceptible of great variation. There is tomato soup thickened with rice; tomato soup seasoned with herbs, or tomato and onion soup. A good plain soup may be made by adding the stewed tomato after it has been strained to remove the seeds, and thickened with flour, cracker crumbs, or a beaten egg, to a kettle of beef broth. Or a broth made from a ham bone may be enriched in this way, and will need no other "trimmings" to render it an excellent, rich soup.

Source: *Good Housekeeping* 13 (September 1891):118.

Tomato Soup (1893)

Ingredients for 2 quarts.

½ tin tomatoes.
1 oz. butter.
Pepper and salt.
2 oz. tapioca.
2 onions.
2 quarts stock or water.

Fry the onions in the butter, add the tomatoes, and after stewing ½ hour, pulp through a sieve, put the water or stock on the fire to boil, then add the tomato and onion.

Next put in the tapioca, which should have been previously soaked for an hour, and stir all till the soup comes to the boil. Boil for ¼ hour, then season with pepper and salt and serve.

Average cost of this quantity, 8*d*.

Source: *Tinned Foods and How to Use Them* (London: Ward, Lock & Bowden, 1893), 19.

Tomato Soup No. 1 (1899)

Take 1 pint of sifted tomatoes, 1 grated onion, 1 tablespoonful of minced parsley, 1 tablespoonful of minced celery, 1 pint of peanut milk or cream as may be desired. Serve hot with crackers or croutons; 1 teaspoonful of sugar and a little salt should be added to the above.[1]

Tomato Soup No. 2 (1899)

Select 6 ripe but sound tomatoes which have a good thick meat. Cut in small pieces, and put them in a stew-pan with 1 pint of rich nut milk, 1 teaspoonful of grated onion, with a sprig of parsley. Cook slowly twenty minutes, and put through a sieve fine enough to remove the tomato seeds; return to the fire and add 1 tablespoonful of corn-starch dissolved in 2 tablespoonfuls of nut cream. Add salt to taste. Serve with crackers or croutons.

Source: Almeda Lambert, *Guide for Nut Cookery* (Battle Creek, Michigan: Joseph Lambert, 1899), 306.

Clear Tomato Soup (1899)

Take 1 pint of stewed tomatoes, 1 teaspoonful of grated onion, 1 tablespoonful of pearl sago, 1 pint of water, 1 pint of thin nut cream, and salt to taste. Cover the sago with the pint of water and let soak for twenty minutes, then stand on the back part of the stove where it will cook very slowly until the sago is transparent and the water almost boiling. Place the tomatoes in a stew-pan, add the parsley and onion, and cook for ten minutes; then add to the tomatoes the nut cream and put through a sieve. Return to the fire, add the salt and lastly the sago, and serve at once. Serve with white-flour crisps or rolls.

Source: Almeda Lambert, *Guide for Nut Cookery* (Battle Creek, Michigan: Joseph Lambert, 1899), 305–306.

Clear Tomato Soup (1899)

"The Columbia"

1 quart of tomatoes.
1 quart white stock or water.

1 dozen cloves.
1 small onion.
1 bunch of parsley.
1 tablespoonful of sugar.
Salt to taste.

Boil together for 1 hour in porcelain or agate kettle. Strain through fine sieve. Put back into kettle and add 1 tablespoonful of cornstarch, rubbed into a smooth paste with cold water, and a piece of butter size of an egg. Season with cayenne or black pepper.

Source: Ladies of the Trinity Parish Guild, comp., *Fredonia Cook Book* (Fredonia, New York: Fredonia Censor Print, 1899), 15.

St. Denis Cream of Tomato Soup (1902)

One quart beef stock, one quart tomato-juice, one pint sweet cream, salt and soda. Heat beef stock, stew and strain tomatoes until you have a quart of juice. Heat and add to stock. Thicken with a little cornstarch. Heat the cream, to which add slowly the tomato soup. Do not let boil after pouring together. Add salt to taste and just before serving add a small pinch of soda. To keep hot, it is best to use a double boiler. The finer the cream, the finer the soup—Mrs. Israel Washburn.

Source: *Good Housekeeping* 35 (August 1902):204.

Tomato Soup Maigre (1902)

Fry to a golden brown a sliced onion,—this is the bottom of a soup pot,—and over it pour a can of tomatoes, which have been well chopped, and two cups of boiling water. When the tomatoes are cooked tender, rub them through a colander, put the soup in the kettle, add a cup of boiled rice, thicken with a tablespoon of butter, rubbed smooth with the same amount of flour, boil up, and serve.

Source: L[ida] Seeley, *Cook Book: A Manual of French and American Cookery* (New York: Macmillan, 1902), 100.

Southern Tomato Soup (1904)

Southern tomato soup is a meal in itself. Wash two quarts of tomatoes, and set over the fire in three pints of water; cook ten minutes, and drain, saving the water for the soup; press the tomatoes through a sieve, add to them one cucumber, peeled and cut small, one large onion sliced, one dozen okras (also sliced), a five cent marrowbone and the water drained from the tomatoes. Simmer for three hours, and just before sending to the table thicken with a tablespoonful of flour wet with cold water. Season with salt, cayenne and three pats of butter.

Source: Celestine Eustis, *Cooking in Old Creole Days* (New York: R. H. Russell, 1904), 7–8.

Tomato Soup (1907)

One can tomatoes, three pints water, 3-4 cup oatmeal, onion chopped, celery or parsley, tablespoonful sugar, salt, pepper, boil three-quarters of an hour, add one-half teaspoonful soda, strain, then add tablespoonful butter, boil up once and serve.

Source: *Kirmess Cook Book: A Collection of Well-tested Recipes from the Best Housekeepers of Jersey City and Elsewhere. Compiled for the Kirmess Given for the Benefit of Christ Hospital of Jersey City, Nov. 7-8-9-10, 1906.* [Jersey City?, c. 1907], 53.

Tomato Soup Tinned (1915)

Material

Tomatoes	1 tin
Sugar	1 teaspoonful
Water (boiling)	1 quart
Rice (boiled)	2 tablespoonfuls
Onions	2
Butter	2 tablespoonfuls
Parsley	a few sprigs
Flour	a little
Salt and Pepper	to taste

Method.—Peel and slice the onions and fry them in half the butter until a light brown, add the tomatoes and stir all together until hot, add the parsley, chopped fine, and pour over the boiling water. Boil gently for half-an-hour and strain, rubbing the tomato through a sieve into the hot liquid, with a wooden spoon. Return to the saucepan, add the sugar, season to taste, bring to the boil, and add the remainder of the butter well rubbed into a little salt. Lastly, put in the rice (boiled), simmer all ten minutes and serve.

Source: *The Everyday Soup Book: A Recipe for Every Day of the Year including February 29th by G. P.* (London: Stanley Paul, 1915), 26.

Tomato Soup (Savoury) (1915)

Tomatoes	1 can
Soda	a pinch
Milk	1½ pints
Edwards' Desiccated Tomato Soup	1 packet
Water or Stock	1 pint

Onion	1
Cloves	4
Bacon	a little
Pepper and Salt	to taste

Method.—Put the tomatoes into a saucepan with the desiccated soup, add the soda, water, and season, to taste. Simmer gently for thirty minutes, stirring occasionally. In another saucepan put the milk, onion (sliced), cloves and the bacon. Stand on one side of the hob and allow to get hot, but not boil. Then rub the tomato soup through a sieve, and return to the pan with the strained milk. Boil up, and serve with toast cut in tiny squares.

Source: *The Everyday Soup Book: A Recipe for Every Day of the Year including February 29th by G. P.* (London: Stanley Paul, 1915), 27.

Essence of Tomato (1924)

To a quart of white or brown soup stock add a cupful and a half of the juice from canned tomatoes and a few grains of sugar. Bring to boiling point, add a little more salt and pepper, if necessary, and strain through cheesecloth.

Source: Ida C. Bailey Allen, *Mrs. Allen on Cooking, Menus, Service* (Garden City, New York: Doubleday, Page, 1924), 128.

Clear Tomato Soup (1924)

1 quart canned or fresh sliced tomatoes
1 pint cold water
½ teaspoonful pickle spice
1 tablespoonful minced onion
1 tablespoonful minced parsley
½ teaspoonful celery seed
½ teaspoonful salt
2 teaspoonfuls sugar
Sliced orange or lemon

Combine all the ingredients, except the orange or lemon, simmer gently for fifteen minutes, strain through a fine sieve, and serve with a thin slice of orange or lemon in each portion. If desired a bit of the grated orange or lemon rind may be added to the vegetables while they are simmering.

Source: Ida C. Bailey Allen, *Mrs. Allen on Cooking, Menus, Service* (Garden City, New York: Doubleday, Page, 1924), 128–129.

TOMATO CONSOMMÉ, BOUILLON, AND BISQUE

Crayfish Bisque a la Créole (1885)

Wash the crayfishes, boil and drain them. Separate the heads from the tails. Clean out some of the heads, allowing two or three heads to each person. Peel the tails. Chop up a part of them, add to them some bread, onions, salt, black pepper and an egg or two. With this dressing, stuff the parts adhering to them. Fry a little garlic, onions, ham, one turnip, one carrot, and a little flour; add some water, the chopped claws, a few tomatoes, thyme, sweet bay, parsley and a little rice, stirring often to avoid scorching. When well boiled, strain through a colander. After straining, put back to the fire and season to taste. Put the stuffed heads into the oven until brown. When ready to serve, put them and the tails aside in a soup dish and pour the soup over them. Before serving, add a little butter and nutmeg, stirring until the butter is melted.

Source: [Lafcadio Hearn], *La Cuisine Creole: A Collection of Culinary Recipes from Leading Chefs and Noted Creole Housewives, Who Have Made New Orleans Famous for Its Cuisine* (New York: Will H. Coleman, 1885), 252.

Crab and Tomato Bisque (1886)

Use for this soup one quart of milk, one pint of cooked and strained tomatoes, half a pint of crab meat, two tablespoonfuls of flour, two of butter, one teaspoonful and a half of salt, half a teaspoonful of pepper, and a scant half teaspoonful of soda.

Put the milk, except half a cupful, on the stove in a double-boiler. Mix the cold milk with the flour, and stir into the boiling milk. Cook for ten minutes.

Put the tomatoes on to cook in a saucepan. When they have been boiling for five minutes, stir in the soda: and continue stirring until there is no froth on the vegetable. Add to the tomatoes the butter, salt and pepper. Add the crab meat to the mixture, stir the contents of the double-boiler into the stewpan. Serve at once.

When it is inconvenient to use fresh crab meat, canned meat may be taken instead.

Source: *Good Housekeeping* 3 (May 29, 1886):43.

Tomato Bisque (1896)

Stew half a can of tomatoes until they become quite soft and will strain readily, then boil a quart of milk in a double boiler. Cook together a tablespoonful of cornstarch and an equal quantity of butter in a small saucepan, adding enough hot milk to make it pour readily, then stir it carefully into the boiling milk, and let it boil for about ten minutes; then add a third of a cup of butter in small lumps, and stir it until it has become well incorporated, add some salt and pepper and the strained tomatoes, and if the tomatoes are very acid, half a saltspoonful of soda may be added. Serve while hot.

Source: Oscar Tschirky, *The Cook Book by "Oscar" of the Waldorf* (Chicago and New York: Werner, 1896), 11–12.

Tomato Bouillon with Oysters (1902)

Cook together one can of tomato and one and a half quarts of bouillon, one chopped onion, one small bay leaf, six cloves, one teaspoon of celery seed and half a teaspoon of peppercorns. Cook for twenty minutes. Strain, cool and clear, then strain into cups over parboiled oysters. For this soup a brown stock would be best suited, but I do not have it today; instead I have veal stock enriched by two teaspoons of beef extract. When the bouillon is cooked, clear it exactly the same way as if you were making a plain clear soup. Beat the white of one egg lightly, just enough to separate it, and add to it the egg shell broken up. When the stock has cooled add this and set it where it will come slowly to the boil, stirring it constantly. The egg will attract all particles of tomato and everything solid. Let it boil two minutes, then strain very carefully through two thicknesses of cheesecloth. It will be perfectly clear, but with the red tomato coloring. If it were left to cool it would become a solid jelly.

While the bouillon is cooking prepare the oysters. I have here a pint of the shellfish. First add to the bowl in which they are one half cup of cold water, in which they must be washed. Stir them slightly in this and pick out each one by itself, rejecting every particle of liquor and put the oysters back in it to parboil. Lift them from the fire as soon as the edges begin to curl. Put a few into each cup, pour the hot bouillon over and serve with croutons.

Source: *Good Housekeeping* 34 (January 1902):51.

Consommé with Tomato Blocks (1902)

For this use, if possible, tomato conserve. If one is, however, obliged to use canned tomatoes take the thickest or best portion of the tomatoes; press through a sieve and then cook slowly to a thick paste. Take a half cup of this paste, add just a suspicion of ground mace, a quarter of a saltspoonful, salt and tabasco. Stir in the whites of two eggs slightly beaten. Pour into a pie dish, stand in a pan of water and cook in the oven until thoroughly set; cut into fancy shapes. Put the consommé in the tureen first, and then carefully drop in the blocks. Serve at once.

Source: Sarah Tyson Rorer, *Mrs. Rorer's New Cook Book* (Philadelphia: Arnold, 1902), 57.

Bouillon a la James Madison (1904)

Two gallons of water, throw in every bone you have (ham bones are excellent), with three good sized carrots, three onions, celery, a can of tomatoes. Salt and pepper pod to taste. Simmer, closely covered, all day and all night. The next morning strain into a large bowl. If in a hurry set the bowl in cold water, otherwise put in cellar or on ice. Remove the grease very carefully. Cut up fine, size of dice, three pounds of rump beef, take two eggs and break them over the cut meat, yolk and white. Stir freely. Add celery, salt and pepper, pour the bouillon on it, settle it on the fire, stir until the froth rises. Skim off very carefully, strain off through a nice clean cloth or flannel. Set aside for use. When ready to serve, warm the quantity desired, throw in small pieces of celery, cover closely, throw a bunch of chervil and a glass of good sherry in the soup according to taste.—*Cook, born in James Madison's family.*

Source: Celestine Eustis, *Cooking in Old Creole Days* (New York: R. H. Russell, 1904), 4–5.

Tomato and Clam Consommé (1905)

To one quart consommé add two cups each, clam water and canned tomatoes. Clear, and add soft part of clams.

To obtain clam water:

Wash thoroughly and scrub two quarts of clams, put in a granite stew pan, add one-half cup cold water, cover closely, place on front range, and let cook until shells open. Remove clams and strain liquor through double cheese cloth.

Source: Fannie Merritt Farmer, *What to Have for Dinner* (New York: Dodge Publishing, 1905), 122.

Consommé Tomatee (1909)

Make about two quarts and a half of consommé from three pounds of veal (knuckle, discarding part of the bone), two pounds of beef (neck) and a fowl. Cook the fowl in the beef and veal broth and reserve it for croquettes or some other dish. Use onion, parsley, carrot, celery, three tomatoes and a soup bag to flavor the soup. Clarify in the usual manner. Flavor with sherry, if desired, and serve in each plate, julienne strips of tomato (no seeds), cooked white of egg, celery and string beans.

Source: Janet M. Hill, "Seasonable Recipes," *Boston Cooking-School Magazine* 14 (October 1909):137–138.

Tomato Bouillon (1910)

Put two cups of boiling water, two cups of tomatoes (cooked or fresh) cut in bits, half an onion, two parsley branches, four slices of carrot and a bit of bay leaf over the fire; let simmer twenty minutes, then drain through a fine sieve. The sieve should be fine enough to hold back the seeds, but the flesh of the tomato (no other vegetable) should pass the sieve. Add a quart of boiling water in which a tablespoonful of beef extract has been dissolved or a quart of meat broth, also salt and pepper as needed. Heat to the boiling point and serve. This soup should be of a bright red color.

Source: Janet M. Hill, "Seasonable Recipes," *Boston Cooking-School Magazine* 14 (March 1910):378.

Simple Tomato Bisque (Soup) (1910)

Scald one quart of milk with a stalk of celery and two slices of onions. Press enough cooked tomatoes through a sieve to make one pint; add half a teaspoonful of salt and pepper as desired. Stir one-third a cup of flour and a teaspoonful of salt with milk to make a smooth batter; dilute with a little of the hot milk, stir until smooth, then stir into the rest of the hot milk. Continue stirring until smooth and thick; cover and let cook fifteen minutes. Strain into the hot purée, mix thoroughly and serve at once with croutons.

Source: Janet M. Hill, "Seasonable Recipes," *Boston Cooking-School Magazine* 15 (April 1910):82.

Clear Tomato Bouillon (1916)

Cook together twenty minutes three pints of broth, free of all fat, one can of tomatoes, one onion, chopped, three branches of parsley, three cloves, two stalks of celery or a few dried celery leaves and half a carrot, cut fine; strain and let cool; beat the whites of three eggs slightly; add the crushed shells and mix through the chilled soup; stir constantly over the fire until boiling; let boil five minutes and draw to a cool place on the range; add one-fourth a cup of cold water and let stand ten or fifteen minutes. Lay a table napkin, wrung out of boiling water, over a colander in a dish, and set a gravy strainer above; pour the soup into the strainer. When all the soup has dripped through, add salt and pepper and reheat.

Source: Janet M. Hill, "Seasonable and Tested Recipes," *American Cookery* 20 (February 1916):529.

Tomato Bouillon, with Oysters (1916)

Cook one can of tomatoes, three pints of beef broth, one-fourth cup of chopped onion, a bit of bay leaf, four cloves, one-fourth cup dried celery leaves, half a green and red pepper pod and three branches of parsley, twenty minutes; strain into a bowl; when cold remove fat and mix with the slightly beaten whites of three eggs and with the crushed shells of several eggs; stir constantly over the fire until the soup boils, let boil three minutes, draw to a cooler part of the stove to settle, and strain through a napkin wrung out of boiling water; add salt as needed and reheat. Serve in bouillon cups with three or four cooked oysters in each cup. To cook the oysters, pour a cup of cold water over a pint of them, wash each oyster in the water, making sure no shell is attached to it; strain the liquid over them, heat over a quick fire, to the boiling point, and let boil two minutes; skim out and use.

Source: Janet M. Hill, "Seasonable and Tested Recipes," *American Cookery* 20 (April 1916):690.

Tomato Bouillon (1917)

Cook one can of tomatoes, two cups of water and half a cup, each, of celery leaves, sliced onion and carrot at a gentle simmer twenty-five minutes, and strain through a napkin wrung out of hot water; reheat with an equal quantity of broth or consommé, three tablespoonfuls of tomato catsup and salt and pepper as needed. This is sometimes served with a spoonful of whipped cream on the top of each cup. The tomato and vegetables, left in the napkin, may be pressed through a fine sieve and used in a dish of rice or macaroni.

Source: Janet Hill, "Seasonable and Tested Recipes," *American Cookery* 21 (June-July 1917):34.

Quick Tomato Bouillon (1924)

5 cupfuls boiling water
3 teaspoonfuls vegetable or meat extract or 3 bouillon cubes
2/3 cupful chili sauce or ½ can tomato soup

Put the extract, or bouillon cubes, whichever is used, into a saucepan; add the boiling water and simmer just long enough to dissolve. Add the sauce or soup, and bring again to boiling point, then strain through a fine sieve.

Source: Ida C. Bailey Allen, *Mrs. Allen on Cooking, Menus, Service* (Garden City, New York: Doubleday, Page, 1924), 129.

Jellied Tomato Cream Bouillon (1924)

Prepare tomato aspic, pour it into a bowl or pan, and when it has begun to congeal add three-fourths cupful of whipped cream, dragging it through the aspic so that it will be streaked in rather than blended.

Source: Ida C. Bailey Allen, *Mrs. Allen on Cooking, Menus, Service* (Garden City, New York: Doubleday, Page, 1924), 132.

COMMERCIAL OR VOLUME TOMATO SOUP

Tomato Soup (1865)

Take a stock-boiler that will hold about twenty gallons. Put into it fifty pounds of beef-shin to fourteen gallons of cold water. Boil it, partly uncovered, for fourteen hours. After the water has partly boiled away add a little hot water from time to time, as it may require. After it has boiled the required time take it from the fire and add to it one quart of cold water. Afterward let it stand for ten minutes. Next, skim off all of the fat and strain the liquor from the meat through a fine sieve, and we shall have very nearly seven gallons of the liquor. Should there be more than seven gallons of the liquor, boil it down to the required quantity, but should there be a less amount, add the difference in hot water. This is called "stock." Next, take one bushel and a half of tomatoes, put them into a boiler, mash them up a little, and let them boil in their own liquor for one hour and a half. Next, strain them through a fine sieve—fine enough to stop the seeds and the skins. All the rest of the tomato must go through the sieve, after which we shall have about six gallons of the tomato liquor. If more than six gallons, boil it down to such amount. If less, add more tomato. Next, mix the stock and the prepared tomato together, and keep the mixture somewhat under a boiling temperature until wanted for further action. Next, prepare the following vegetables: Peel and weigh one pound and a half of onions, the same amount of turnips, one pound and three-quarters of carrots, and one pound of beets. Chop them all together quite fine. Next, take a soup-boiler that will hold sixteen gallons. Put into it three and a half pounds of butter. Next, add the chopped vegetables. Put the boiler on a hot fire, and cook the vegetables well. Next, add to them three and one-quarter pounds of flour, and thoroughly mix the whole together while hot. Next, take the boiler from the fire and let it cool a little. Next, add one ounce of black pepper, one-half a pound of brown sugar. Mix the whole well together, and add the mixture of beef-stock and tomato. The composition must now be well stirred for about ten minutes, and afterward put on the fire and stirred until it may boil. Continue to let it boil and skim it for about five minutes, after which strain it through a fine sieve, but do not press the vegetables through the sieve. The composition will then be ready for the table, or for being hermetically sealed in cans. The amount of the preparation (which I term "tomato Soup") so made will be about thirteen gallons. It is a composition containing preservative qualities, which will prevent it from decomposition for a great length of time.

Source: James H. W. Huckins, U.S. Patent no. 47,545, issued May 2, 1865.

Tomato Soup (for 60 men) (1910)

Ingredients used:

3 gallons tomatoes, or
8 No. 3 cans tomatoes.
3 gallons beef stock.
1 pound bacon.

Mix all ingredients well and boil for one and one-half hours; remove the bacon and press the soup through a colander to separate the skins and seeds of the tomatoes. Replace on the range and thicken slightly with a flour batter; pepper and salt to taste and color lightly with brown sugar. Regulate the amount of beef stock so that, when the soup is ready to serve, there will be about 6 gallons.

Over-ripe or bruised tomatoes may—to prevent waste—be used in the preparation of this soup.

Source: *Manual for Army Cooks* (Washington, D.C.: Government Printing Office, 1910), 68.

Tomato Soup (1914)

84 gallons Pulp from machine, or its equivalent in canned pulp.
28 gallons Water.
12 pounds Salt.
12 pounds Butter and 11 pounds Sugar.
6 pounds Flour.
10 pounds Chopped Onions.
½ pound Chopped Garlic.
6 ounces Ground White Pepper.
6 ounces Bay Leaves.
1 ounce Powdered Saigon Cinnamon.

Place the butter and pulp in kettle and bring to boil; then add the onions and garlic and simmer one hour, adding water to make up for loss by evaporation. When the cook is half finished, add the bay leaves, and when within ten minutes before the time is up add the cinnamon and the flour mixed with water; boil two minutes, turn off steam, run through rotary pulper or shaker, place in cans, seal and process. No. 3 cans 45 minutes at 240 degrees.

Source: Canning Trade, *A Complete Course in Canning,* 3rd ed. (Baltimore: Canning Trade, 1914), 155.

Condensed Tomato Soup (1914)

120 gallons Pulp from machine or its equivalent in canned pulp.
1 gallon Water (cold).
14 pounds Salt.
12 pounds Sugar.
15 pounds Onions, chopped fine.
2 pounds Butter.
1 pound Beef Extract.
5 pounds Corn Starch.
6½ ounces Ground White Pepper.
1 ounce Ground Saigon Cinnamon.
½ ounce Mace.

Place the pulp and butter in a kettle and bring to a boil, then add the onions and cook 40 minutes. When the cook is within 10 minutes of the finish, add the extract of beef, salt and sugar and pepper; when within 2 minutes, add the cinnamon and mace; mix the cold water and corn starch, stir in, and when it boils, turn off steam; then run through rotary pulper or shaker, place in cans, seal, and process No. 1 cans 30 minutes at 240 degrees.

Source: Canning Trade, *A Complete Course in Canning,* 3rd ed. (Baltimore: Canning Trade, 1914), 155–156.

RECIPES FOR CANNING AND PRESERVING TOMATOES

Preserved Tomatoes (1820)

1 lb large size Tomatoes—1 lb best Brown sugar—1 large lemon—grade the rind into a dish, squeese the juice into a bowl, Scald the tomatoes and peel off the skin, boil the whole slowly 1½ hours, skimming them carefully and then add the lemon juice and boil the whole ½ hour longer.

Source: Dr. Benjamin F. Heyward, from Worcester, Massachusetts, Recipe Book, purportedly written in 1820. Handwritten manuscript in the Old Sturbridge Village Library, p. 16.

To Preserve Tomatoes through the Winter (1822)

Peel the Tomatoes, cut them small, and stew them without water, their own juice being sufficient: season them with salt, pepper, grated ginger, garlic pounded fine, to your taste—when cool, put them up in bottles and cork them so as to exclude the air—look at them frequently, if you observe an effervescence of mould, or a disposition to foment—heat them over a slow fire—they must be done in an earthen pan, or the fine red colour will not be so well preserved—they require to be kept on the fire some considerable time, until some are wasted, or they will not keep—when the weather is cool there will be no further trouble with them.

Source: *American Farmer* 4 (September 27, 1822):208.

Preserved Tomatoes (1832)

Dining, a few days since, at Mr. Pardoe's, innkeeper, three miles below Lewisburgh bridge, my attention was called to notice a new kind of preserve, prepared by the landlady from the common tomato. Its flavor was remarkably rich and fine, so much so, that I was induced into the particulars. It is a discovery of her own. The tomato is taken when nearly ripe, and prepared in the usual manner of other preserves, with sugar and molasses. When Mrs. Pardoe found that the experiment succeeded beyond her expectations, she would have cured more of them, but that the frost had injured what remained in her garden as to unfit them for being used.

Source: *Pennsylvania Reporter,* November 23, 1832.

To Preserve the Tomato (1835)

Press the ripe pulp through a hair sieve—spread it thin on plates or tins so that it may become quite dry, or it may be dried in a moderately heated oven. A small piece of this dried pulp will flavour a Tureen of Soup.

Source: *Pennsylvania Inquirer and Daily Courier,* August 20, 1835, 1.

Preserved Tomatas (1838)

Take large fine tomatas, (not too ripe,) and scald them to make the skins come off easily. Weigh them, and to each pound allow a pound of the best brown sugar, and the grated peel of a large lemon. Put all together into a preserving kettle, and having boiled it slowly for three hours, (skimming it carefully,) add the juice of the lemons, and boil it an hour longer. Then put the whole into jars, and when cool cover and tie them up closely. This is a cheap and excellent sweetmeat; but the lemon must on no account be omitted. It may be improved by boiling a little ginger with the other ingredients.

Source: Eliza Leslie, *Directions for* Cookery, 3rd ed. (Philadelphia: E. L. Carey & A. Hart, 1838), 441.

To Preserve Tomatoes (1839)

To preserve tomatoes in sugar, take them when ripe, but not too soft, skin them, cut them in two, taking out the seeds—take for one pound of tomatoes three-quarters of a pound of sugar, loaf or brown; loaf is best to keep them—First take two fresh lemons, cut them in thin slices, put them in one quart of water, boil them until soft, strain the seeds out, add the sugar, let it boil for half an hour slowly, then put in as many tomatoes as not to crowd them. Let them boil until clear, then take them out carefully with a skimmer, lay them on a large dish, then put more in and boil them the same; do so until all are done. When cold put them in glass jars—If there is not syrup enough to cover them, make a little more. When all is cold, dip white writing paper in brandy, cover them with it, put double paper over them, tie them tight, and keep them in a cool dry place.

Source: *American Farmer* 3rd ser. 1 (August 28, 1839):110.

Canning Tomatoes—Tin Cans—Soldering (1867)

A Housekeeper in the last number of the Co. Gent., inquires about Canning Tomatoes. According to my observation, many others experience the same difficulty in keeping them in glass. Some use stone jars, crocks, &c., and think they succeed pretty well, but I find it rather common to meet with tomatoes with some traces of fermentation; and rather uncommon to meet with those, put up in the manner most in vogue with families in the country, that are entirely sweet and good as when used fresh from the vines.

If the use of Mason's fruit jar, according to directions, will almost invariably insure perfect success with tomatoes, the following instructions are, perhaps, unnecessary, but my experience has led me decidedly to prefer tin rather than glass.

Last season I filled one dozen glass jars from a kettle of the hot tomatoes, and scalded them carefully. The remainder were put in tin cans and soldered up. The scalding of the jars appeared to be perfect, but the contents, I believe, of every one spoiled. Those put in the cans were mixed with a larger lot, and therefore lost sight of to some extent, but I have no reason to suppose but that they kept in good order with only the usual percentage of loss, as they were treated precisely the same as some thousands of others that did. I tried the same experiment again with additional care, but with the same result. Perhaps if they had been very thoroughly cooked and well seasoned, the result might have been different—but this thorough cooking and high seasoning, I think, destroyed, to some extent, that perfect freshness with which we like all canned fruit to open out.

Now, if housekeepers would save their glass jars for peaches, &c., and for tomatoes use only tin cans, they need have but little difficulty in always having fresh tomatoes, just as good for cooking as the day they were taken from the vines.

Be careful to get good cans. A well made fruit can has the top and bottom soldered on from the inside, and the small space between the turned-up edge of the top and bottom, and the side of the can is *entirely* filled with solder. If there are places here and there where this space is not thus filled up, the can is not as reliable as it should be. They should also have a small gutter around the edge of the hole in the top, for the cap to fit in; this gutter makes it much easier to solder on the cap than when the latter fits flatly on the top. After scalding and peeling the tomatoes, throw them into a kettle, (an iron one is as good as any if it be entirely free from rust,) bring to a boil, and allow them to remain for about five minutes; then take the kettle off the fire, fill up the cans and solder on the caps.

For those entirely unaccustomed to using a soldering iron, some directions may be necessary. In selecting an iron, get a large-sized one; it holds the heat better than the small one. Two irons are better than one if many cans are to be soldered, as one can be heating while the other is in use. A charcoal furnace, such as tinners use for heating the irons, is also a convenience, but for canning a small number for family use, the kitchen fire will answer. To put the iron in working order, smooth with a file the four sides of it towards the point, and then rub it on a board with solder and plenty of rosin, until the solder adheres and makes a bright surface. This smooth, bright surface is necessary to make it do good work.

In filling up the cans, if any of the juice is spilled into the gutter, wipe it out with a dry cloth. Having adjusted the cap, sprinkle a little powdered rosin around the edges; then taking the hot iron, drop three or four little globules of melted solder around the gutter, and pass the iron around till all is fast. The heat of the iron, together with the hot tomato, will cause a considerable escape of steam during this process, and if it has to force its way out under the edge of the cap, it will be difficult to make the solder stick. It is therefore best to punch a small hole in the centre of the cap for the escape of the steam. This hole can be stopped by a drop of solder after all is tight around the edge.

As soon as the cans are cold, if they are air tight, the bottom will be pressed in. If at any time during the season they are on hand, the bottom begins to bulge out, the sooner the contents are used the better, for they are beginning to spoil. Freezing will, however, give them that appearance, but in that case they will be all right again as soon as thawed. With proper care, and after allowing for breakage of glass, there is but little more expense, if any, attending the use of tin cans than of glass jars.

The soldering iron will last a life time with good usage, and no more service than it would have in most families, and is useful for many purposes besides soldering and unsoldering fruit cans. The cans will last for years if they are always carefully opened with the hot iron, and immediately after being emptied are rinsed and dried. I know a housekeeper who has had them in use for five or six years, and they are still in good order and ready for further service.

Tomatoes seem to be fast becoming, if they have not already become, a staple article of food all the year round; and with little other outlay than what is necessary to purchase a sufficient number of cans to start with, and a soldering iron or two, most families in the country may have a continued and abundant supply—T. W., *Cecil Co., Md., 7 mo. 31st, 1866.*

Source: *Country Gentleman* 28 (August 1, 1866):115.

Canning Tomatoes (1868)

For canning tomatoes, take ripe tomatoes and strip the skins off; cut in thick slices, lay them in a pan, season with salt, pepper and butter to your taste, cook until tender, and then place them carefully in your cans. I use "Mason's" and "Spencer's" fruit cans only.—*Prattsburgh, N.Y., Mrs. W.*

Source: *Country Gentleman* 31 (May 28, 1868):393.

Canned Tomatoes (1869)

Scald and skin the tomatoes; drain off all the juice. Fill the can till it weighs two pounds. Add one large-spoonful of syrup made in the following manner: To one gallon of water one and one half pound of salt; the same of sugar. Seal the cans; boil thirty minutes. Take them from the water; open the vent for the escape of the gas, and seal again.

Another Receipt for Cooking Tomatoes (1869)

Cook the tomatoes in an enameled kettle ten to fifteen minutes. Take the bottles from the hot water; fill them. Use the same allowance of syrup to a quart; cork and seal while the bottles are hot.

Source: Jane Croly, *Jennie June's American Cookery* (New York: American News, 1869), 345–346.

Canning Tomatoes (1875)

Seeing an inquiry in your paper on this subject, I send you a receipt which I have found to be good:

For a beginning, I use tin cans, from the fact I think the fruit should be kept from the light altogether after being canned, and I think also that they keep better in tin. I never lost but one can in my life, and that was the first time I ever attempted to can them. I do not think now it was the can's fault. I have never tried glass or stone, but my neighbors have, and they lose more than they keep, and some lose all, while mine, so far, have kept perfectly good. We buy prepared sealing-wax. Gather the tomatoes, scald, skin and slice them; put them in a brass kettle, which must be perfectly bright and free from canker; this can be done quickly by scouring with vinegar and fine sand, ashes or salt, being careful not to have anything in it that will scratch your kettle. Put the tomatoes in and let them come to a boil, so as to heat them entirely through. Have a pot (that will come almost to the top of your cans) boiling all the time you are canning, to set your cans in while filling and sealing; this will exclude all the air. Now fill the cans full, not nearly full, but level full, keeping them boiling in water until they are full, and then put on the top tight, and seal with your wax warm enough to run well, but not hot enough to run into the can; lift the can out of the water and set where a draft of air cannot strike them until cool. Now, to be sure they are air tight, when cool melt some more wax and run around your cans again. This will stop all airholes, if any. During warm weather keep your cans in a dry, cool room, or closet. I prefer a dark place for keeping them, if possible; and when the hard winter comes keep them where they will not freeze if you can, as freezing will crack the wax and let in the air. Mine froze the top off on the floor last winter during the cold snap, and I recanned them and never lost a can. I have tried to be explicit in giving my experience in canning tomatoes.

Source: *Prairie Farmer* 46 (October 2, 1875):315.

To Can Tomatoes (1876)

Two and three pound cans are generally used. The tomatoes must be full ripe and smooth. We should recommend the variety known as Boston market, as the best, if true to name. If desirable to pack in large quantities, a 25 gallon farmer's kettle should be used for scalding, and a large tub for cooling. The water in the cooling tub should be kept cool either by a running stream of cool water, or ice. Provide a two or four quart dipper perforated with ½ inch holes, made of strong tin, or sheet iron. The dipper should have a long wooden handle. Have the water boiling in the kettle; wash the tomatoes clean and put them in the perforated dipper; hold in the boiling water until the skin loosens, then lift out, drain and hold in the cold water, shaking the dipper the while, the skin will end up and nearly come off; after draining put in pans and send to the peelers; have the remainder of the skin and any other imperfections carefully removed. Great care should be taken not to allow any unripe particles of the tomatoes in the cans. Pack solid, keeping the tomatoes as whole as may be, drawing off the juice and making the meat as solid in the can as possible. After the cans are full, weigh. They should weigh as near as possible, the weight the can indicates. After weighing, they should be set in a bath, or kettle, with the water just far enough up the sides of the can, so as not to overflow when boiling. Several plans are adopted for this purpose, but all that is necessary to say here is, the can should be made boiling hot, without any water or steam mingling with the tomato; when hot take from the bath, see that the cans are full, seal up tight, boil hard for 30 minutes, and cool. The same caution is necessary here as to tight, clean cans, and regularity about weighing and processing.

Source: William Archdeacon, *The Kitchen Cabinet: A Book of Receipts* (Chicago: Published for the author, 1876), 143–144.

To Can Tomatoes—Old Style (1876)

Scald and peel the tomatoes as in previous formula; put in a kettle, stir slowly until boiling hot; break the tomato by stirring as little as possible. Provide a wide mouth funnel the size of the hole in the can; fill up, keeping the tomato as whole as possible, and having the whole very thick; do not allow much of the thin juice of the tomato to go into the can; seal tight while hot, and process 20 minutes; cool. Process 5 minutes less for second class.

Source: William Archdeacon, *The Kitchen Cabinet: A Book of Receipts* (Chicago: Published for the author, 1876), 144–145.

To Can Tomatoes (1876)

Let them be entirely fresh. Put scalding water over them to aid in removing the skins. When the cans with their covers are in readiness upon the table, the red sealing-wax (which is generally too brittle, and requires a little lard melted with it) is in a cup at the back of the fire, the tea-kettle is full of boiling water, and the tomatoes are all skinned, we are ready to begin the canning. First put four cans (if there are two persons, three if only one person) on the hearth in front of the fire; fill them with boiling water. Put enough tomatoes in a porcelain preserving kettle to fill these cans; add no water to them. With a good fire let them come to the boiling-point, or let them all be scalded through. Then, emptying the hot water from the cans, fill them with hot tomatoes; wipe off the moisture from the tops with a soft cloth, and press the covers on tightly. While pressing each cover down closely with a knife, pour carefully around it the hot sealing-wax from the tin cup, so bent at the edge that the wax may run out in a small stream. Hold the knife still for a moment longer, that the wax may set. When these cans are sealed, continue the operation until all the tomatoes are canned. Now put the blade of an old knife in the coals, and when it is red-hot run it over the tops of the sealing wax to melt any bubbles that may have formed; then, examining each can, notice if there is any hissing noise, which will indicate a want of tightness in the can, which allows steam to escape. If any holes are found, wipe them, and cover them while the cans are hot with a bit of the sealing-wax. There will be juice left after the tomatoes are canned. Season this and boil it down for catchup.

Source: Mary Henderson, *Practical Cooking and Dinner Giving* (New York: Harper & Brothers, 1876), 245–246.

Canning (1882)

The tomatoes first undergo a scalding process, before being taken into the factory in wooden pails. Here they are handed to women for paring. The women and girls are seated around tables, and at five cents a pail they can earn fair wages. The tomatoes are next inspected to detect any unpeeled tomatoes. Then they are passed through a funnel-shaped machine into the cans, after having been cut in the passage. The contents of the cans are slightly compressed and a portion of the juice poured off. The top of the can is then quickly soldered on by a workman, who uses a heavy iron, heated by gasoline for the purpose, or by an apparatus called a capping machine. The canned tomatoes are now loaded on trays, and conveyed to the "bath" room, where they are placed in vats of water heated by steam, and slowly cooked. After this they are allowed to cool, carried to the storehouse and labeled.

Source: Artemas Ward, *The Grocers' Hand-Book and Directory* (Philadelphia: Philadelphia Grocer Publishing, 1882), 265.

BOUILLABAISSE

Bouillabaisse à la Marseillaise (1840)

This dish is composed of whiting, red mullet, tench, turbot, eels, or any other fish one might have. The greater the variety of fish, the better the bouillon will be; for it concerns a potage highly esteemed by the Provençals. Scale and gut your fish, clean them well, then slice them into chunks, cross-wise; throw a good handful of sliced onions in a very large pot, with [olive] oil; lightly sauté the onion, without allowing it to color; then add your fish, with a bundle of herbs, whole cloves, whole black pepper, salt, cloves of garlic, plenty but not too much, and lots of tomatoes [love apples], cut into small chunks which you scatter about between layers of the fish. Anoint with more [olive] oil, and fill your pot with water. Not only must the fish bathe in water, but there must be sufficient liquid to result in a proper potage. Cover the pot, and cook for an hour. Taste your bouillon, in order to correct the seasoning, observing that it be a little high, because the fish tends to tone it down. When done, line the tureen with slices of bread, pour some of the bouillon over so that the bread soaks it up, then arrange the fish on top; cover with more slices of bread, pour on the rest of the bouillon, and serve.

Source: Mme. Utrecht-Friedel, *La petite cuisinière habile . . .* (New Orleans: n.p., 1840), 64–65, translated from the French by Karen Hess.

Bouillabaisse Soup (1854)

Put the following ingredients into a saucepan to boil on the fire:—four onions and six tomatoes, or red love-apples, cut in thin slices, some thyme and winter savory, a little salad oil, a wine-glassful of vinegar, pepper and salt, and a pint of water to each person. When the soup has boiled fifteen minutes, throw in your fish, cut in pieces or slices, and, as soon as the fish is done, eat the soup with some crusts of bread or toast in it. All kinds of fish suit this purpose.

Source: Charles Elmé Francatelli, *A Plain Cookery Book for the Working Classes*, new ed. (London: Routledge, Warne, and Routledge, 1852), 63–64.

Bouille-abaisse (1884)

Take six pounds of codfish; cut it up into small pieces; chop two red onions; put them in a stewpan with an ounce of butter; let them brown without burning. Now add the fish and four tablespoonfuls of fine olive-oil, a bruised clove of garlic, two bay leaves, four slices of lemon peeled and quartered, half a pint of Shrewsbury tomato catsup, and half a salt-spoonful of saffron. Add sufficient hot soup stock to cover the whole; boil slowly for half to three-quarters of an hour; skim carefully while boiling; when ready to serve add a tablespoonful of chopped celery tops.

Source: Thomas J. Murrey, *Fifty Soups* (New York: White, Stokes & Allen, 1884), 15–16.

Bouillabaisse (1884)

Take one black fish, one sea-bass, a small striped bass, a medium-sized live lobster, an eel, and two soft-shell crabs: clean them all and cut them in pieces; put in a thin saucepan, pretty large and not too deep, two ounces of sweet oil, two cloves of garlic, and a bay-leaf; fry a little, then put the fish in the saucepan with a dozen large raw oysters; season with salt and pepper, a sliced lemon, three tomatoes peeled and cut small, half a teaspoonful of powdered saffron, a pint of white wine, and cold water enough to cover the fish; put this on a very brisk fire, and in fifteen minutes all the fish ought to be done and the gravy sufficiently reduced; add a tablespoonful of chopped parsley and boil a minute longer; put slightly toasted slices of bread in a deep dish, pour the liquid over, and send to table with the fish in another dish. At table, serve the same plate from the two dishes.

Source: Felix Déliée, *The Franco-American Cookery Book* (New York: G. P. Putnam's Sons, 1884), 458.

Bouillabaisse (1885)

Chop some onions and garlic very fine, fry them in olive oil, and when slightly colored add some fish cut up in slices; also a few tomatoes scalded, peeled and sliced, some salt, black and red pepper, thyme, sweet-bay, parsley, and half a bottle of white wine, and enough water to cover the fish. Put it over a brisk fire and boil a quarter of an hour. Put slices of toasted bread in a deep dish, place the fish on a shallow dish with some of the broth, and pour the balance on the bread and serve hot.

Source: [Lafcadio Hearn], *La Cuisine Creole: A Collection of Culinary Recipes from Leading Chefs and Noted Creole Housewives, Who Have Made New Orleans Famous for Its Cuisine* (New York: Will H. Coleman, 1885), 252.

Bouillabaisse (1894)

Prepare one-half pound of red snapper, one-half pound of lobster, one-half pound of perch, one-half pound of sea bass, one-half pound of black-fish, one-half pound of sheepshead, one-half pound of cod, one-half pound of mackerel. Cut all of these fish into two and a quarter inch squares. Mince a fine Julienne of a quarter of a pound of carrots, two ounces of chopped onions, and two ounces of leeks; have also two cloves of garlic. Heat in a saucepan, one gill of sweet oil, add to it the vegetables and garlic, fry them without allowing to color, then add one tablespoonful of flour, mix altogether, and put in the fish; moisten to its height with half white wine and half water, adding two medium sized peeled tomatoes, cut in two, pressed out, and chopped up coarsely; let boil for fifteen minutes on a quick fire to reduce the moistening, then add one clove of crushed and chopped garlic, some saffron, salt, pepper, the pulp of a lemon pared to the quick, and chopped parsley; remove the two whole cloves of garlic. Serve the broth or stock in a soup tureen, the fish separately as well as thin slices of toasted bread; should the moistening not be sufficient for the soup, then add some fish broth to it.

Source: Charles Ranhofer, *The Epicurean* (New York: R. Ranhofer, 1894), 264.

Bouillabaisse (1901)

6 Slices of Red Snapper.
6 Slices of Redfish.
½ Bottle of White Wine.
½ Lemon.
6 Large Fresh Tomatoes, or ½ Can.
3 Onions.
1 Herb Bouquet.
3 Cloves of Garlic.
3 Bay Leaves.
3 Sprigs of Thyme.
3 Sprigs of Parsley.
6 Allspices.
2 Tablespoonfuls of Olive Oil.
1 Good Strong Pinch of Saffron.
Salt, Pepper and Cayenne to Taste.

First cut off the head of the Red Snapper and boil it in about one and a half quarts of water, so as to make a fish stock. Put one sliced onion and a herb bouquet consisting of thyme, bay leaf and parsley, into the water. When reduced to one pint, take out the head of the fish and the herb bouquet and strain the water and set it aside for use later on.

Take six slices of Redfish and six slices of Red Snapper of equal sizes and rub well with salt and pepper. Mince three sprigs of thyme, three sprigs of parsley, three bay leaves and three cloves of garlic, very, very fine, and take six allspice and grind them very fine, and mix thoroughly with the minced herbs and garlic. Then take each slice of fish and rub well with this mixture till every portion is permeated by the herbs, spice and garlic. They must be, as it were, soaked into the flesh, if you would achieve the success of this dish. Take two tablespoonfuls of fine olive oil and put into a very large pan, so large that each slice of the fish may be put in without one piece overlapping the other. Chop two onions very fine and add them to the heating oil. Lay the fish slice by slice in the pan and cover, and let them "étouffé," or smother for about ten minutes, turning once over so that each side may cook partly. Then take the fish out of the pan and set the slices in a dish. Pour a half bottle of white wine into the pan and stir well. Add a half can of tomatoes or six large fresh tomatoes sliced fine, and let them boil well. Then add half a lemon, cut very thin slices, pour over a pint of the liquor in which the head of the snapper was boiled. Season well to taste with salt, pepper and a dash of Cayenne. Let it boil till very strong and until reduced almost one half; then lay the fish slice by slice, apart one from the other, in the pan, and let boil five minutes. In the meantime have prepared one good pinch of saffron, chopped very fine. Set it in a small deep dish and add a little of the sauce in which the fish is boiling to dissolve well. When well melted and when the fish has been just five minutes in the pan, spread the saffron over the top of the fish. Take out of the pan, lay each slice on toast, which has been fried in butter; pour the sauce over and serve hot immediately. You will have a dish that Lucullus would have envied.

Source: *The Picayune Creole Cook Book,* 2nd ed. (New Orleans: Picayune, 1901), 44–45.

Bouillabaisse (1902)

1 rock fish
¼ cup of olive oil
1 level teaspoonful of salt
1 teaspoonful of salt
1 teaspoonful of pepper
1 bay leaf
1 pint of white wine
1 large onion
3 cloves of garlic
2 good-sized tomatoes
1 teaspoonful of powdered thyme
1 tablespoonful of chopped parsley
A dash of cayenne

Peel the onion and garlic, and cut them into thin slices; put the oil into a large saucepan, when hot add the onion and garlic, and stir and cook until browned; then add the fish, cut into slices crosswise. Scald the tomatoes, peel, and cut them into halves; press out the seeds, and cut the flesh of the tomatoes into small pieces. Put them over the fish, pour over the wine, and add sufficient water to just cover the fish. Cover the saucepan, put over a quick fire, and boil thirty minutes. While this is boiling, remove the crusts from six thick slices of bread; toast them in the oven until crisp and brown. Then put them into a deep vegetable dish. Remove the fish with the skimmer and arrange it neatly in the centre of a platter; drain a portion of the broth over the bread, and the balance over the fish. Garnish with chopped parsley, and send to the table.

Source: Sarah Tyson Rorer, *Mrs. Rorer's New Cook Book* (Philadelphia: Arnold, 1902), 685–686.

Bouillabaisse (1917)

Bouillabaisse is only to be made of the many rock fish of curious forms and gorgeous coloring, freshly caught, which live about the lovely rocky calanques of the Mediterranean. To *thon, dorade, mulet, rouget* and *rascasse* (the names of which you will admit sound like those of no other of the finny tribe extant) are added crayfish and *langouste.* After being boiled, the fish mixture is steeped slowly in a rich juice of virgin olive oil and tomato purée and flavored with a liberal dash of powdered saffron and a leaf or two of fennel until it becomes a brilliant, not to say pungent yellow.

Source: Blanche McManus, "Pedigree of the American Boiled Dinner," *American Cookery* 21 (February 1917):516.

CHOWDER

Chowder (1867)

This popular dish is made in a hundred different ways, but the result is about the same.

It is generally admitted that the boatmen prepare it better than others, and the receipts we give below come from the most experienced chowder-men of the Harlem River.

Potatoes and crackers are used in different proportions, the more used, the thicker the chowder will be.

Put in a *pot* (technical name) some small slices of fat salt pork, enough to line the bottom of it; on that, a layer of potatoes, cut in small pieces; on the potatoes, a layer of chopped onions; on the onions, a layer of tomatoes, in slices, or canned tomatoes; on the latter a layer of clams, whole or chopped (they are generally chopped), then a layer of crackers.

Then, repeat the process, that is, another layer of potatoes on that of the clams; on this, one of onions, etc., till the pot is nearly full. Every layer is seasoned with salt and pepper. Other spices are sometimes added according to taste; such as thyme, cloves, bay-leaves, and tarragon.

When the whole is in, cover with water, set on a slow fire, and when nearly done, stir in gently, finish cooking, and serve.

Source: Pierre Blot, *Hand-Book of Practical Cookery* (New York: D. Appleton, 1867), 159–160.

Fish Chowder. No. 2 (1878)

Take a red or any other firm fish, cut it in pieces about three inches square; one pound and a half of salt pork cut [i]n thin slices; one dozen and a half of Irish potatoes, and the same of tomatoes, both sliced thin; half a dozen onions cut fine, and one dozen hard crackers broken in small pieces. Take a large pot, put a layer of pork on the bottom, then a layer each of fish, potatoes, tomatoes, onions and crackers. Sprinkle each layer with a little salt, black pepper, and flour. Repeat the layers of pork, fish, potatoes, etc., until all are used. Fill with hot water until it covers the whole. Put the pot on the fire, and let it boil thirty minutes; then add a pint of claret and boil five minutes longer. The chowder will then be ready for the table.

Source: Ladies of the Saint Francis Street Methodist Episcopal Church, South, Mobile, Alabama, comp., *Gulf City Cook Book* (Dayton, Ohio: United Brethren Publishing House, 1878), 20.

Danbury Clam Chowder (1880)

Use for six persons one quart of clams, one pint of canned tomatoes (or one quart of fresh tomatoes), one quart of sliced potatoes, one pint of sliced onions, one pint of water, half a teaspoonful each of powdered thyme, summer savory, and sweet-marjoram, half a teaspoonful of celery seeds (or, when it is convenient to get fresh celery, use half a pint of it, chopped fine), one quarter of a pound of salt pork, one teaspoonful of pepper, and three teaspoonfuls of salt.

Place a colander in a basin, and turn the clams into it. Now pour the water over the clams, stirring well with a spoon in order to make the washing a thorough one. Save the clam liquor and water that fall into the basin. Chop the clams rather fine, and put them in a cool place until the time for cooking.

Cut the pork in thin slices, and fry slowly for ten minutes. Add the sliced onion to the pork, and cook on a hotter part of the stove for ten minutes, stirring frequently. The onions should be tender, but not browned, at the end of the ten minutes. Turn the pork and onions into a stew-pan, and add to them the clam juice and water, the potatoes, tomatoes, and the celery, if that is to be used. Cook for thirty minutes; then add the seasoning and chopped clams, and cook ten minutes longer. Taste, to be sure that there is seasoning enough; and if no seasoning is required, serve the chowder.

This is a very savory dish. When it is served in a dinner, it is well to have the meat or fish in the next course simple and light.

Source: Maria Parloa, *Miss Parloa's Kitchen Companion* (Boston: Dana Estes, 1880), 161–162.

Clam Chowder (1884)

Chop up fifty large clams; cut eight medium sized potatoes into small square pieces, and keep them in cold water until wanted.

Chop one large, red onion fine, and cut up half a pound of larding pork into small pieces.

Procure an iron pot, and see that it is very clean and free from rust; set it on the range, and when very hot, throw the pieces of pork into it, fry them brown; next add the onion, and fry it brown; add one fourth of the chopped potato, and two pilot crackers quartered, a teaspoonful of salt, one chopped, long, red pepper, a teaspoonful of powdered thyme and half a pint of canned tomato pulp. Repeat this process until the clams and potato are used, omitting the seasoning; add hot water enough to cover all, simmer slowly three hours. Should it become too thick, add more hot water; occasionally remove the pot from the range, take hold of the handle, and twist the pot round several times; this is done to prevent the chowder from burning. On no account disturb the chowder with a spoon or ladle until done; now taste for seasoning, as it is much easier to season properly after the chowder is cooked than before. A few celery tops may be added if desired.

Source: Thomas J. Murrey, *Fifty Soups* (New York: White, Stokes & Allen, 1884), 19–20.

Clam Chowder (1886)

50 clams
1 pound of veal
½ pound bacon or ham
1 pint of stewed or canned tomatoes
1 pint of water
1 pint of milk
6 water crackers or three sea biscuit
1 teaspoonful of thyme
1 teaspoonful of sweet marjoram
1 tablespoonful of chopped parsley
3 medium-sized potatoes
Salt and pepper to taste

Line the bottom of the saucepan with the bacon or ham cut into dice. Pare and cut the potatoes into dice. Chop the onion fine. Cut the veal into pieces a half-inch square. Chop the clams. Mash the crackers. Now put a layer of the potatoes on the bacon or ham, and then a sprinkling of onion, thyme, sweet marjoram, parsley, salt and pepper, and then a layer of veal, then tomatoes, then a layer of chopped clams, and continue these alternations until it is all in, having the last layer clams. Now add the water, which should be boiling and barely cover the whole. Cover closely, place on a slow fire and *simmer* for half an hour without stirring. Then add the milk and crackers, stir and cook ten minutes longer and serve very hot.

The tomatoes may be omitted if not liked.

Source: Sarah Tyson Rorer, *The Philadelphia Cook Book* (Philadelphia: Arnold, 1886), 34.

Fish Chowder (1886)

3 pounds of fish
1 pint of milk
3 medium-sized potatoes
1 quart of water
1 pint of stewed or canned tomatoes
1 large-sized onion
¼ pound of bacon or ham
1 tablespoonful of thyme
1 teaspoonful of sweet marjoram
6 water crackers or three sea biscuit
Salt and pepper to taste

Cut the fish, the potatoes, the onion, and bacon or ham, into pieces about a half-inch square. Now put the bacon or ham and the onion into a large frying-pan, stir and fry them a light brown. Put a layer of the potatoes in a sauce-pan, then a layer of fish, then a sprinkling of onions and bacon or ham, then a layer of tomatoes, then a sprinkling of thyme and sweet marjoram, salt and pepper, and continue these alternations until all is in, having the last layer the potatoes. Now add the water. Cover closely, place over a moderate fire and let it *simmer* twenty minutes without stirring. In the meantime put the milk in a farina boiler and break into it the crackers; let it stand three minutes. Now add this to the chowder, stir, let it boil once, see that it is properly seasoned and serve very hot.

The tomatoes may be omitted if not liked.

Source: Sarah Tyson Rorer, *The Philadelphia Cook Book* (Philadelphia: Arnold, 1886), 35–36.

Clam Chowder (1888)

Chop four dozen clams, peel and slice ten raw potatoes, cut into dice six tomatoes (if they are canned a coffeecupful), add one pound soda crackers, put the pork in the bottom of pot and try it out, partially cook the onion in pork fat, remove the mass from the pot and put on a plate bottom-side up, make layers of the ingredients, season with salt and pepper, cover with water, cook one and one-half hours, adding chopped parsley to taste.

Source: Ladies' Aid Society of the Church of Our Father (Universalist), *Helps to Housekeeping* (Chicago: Commercial Printing Company, 1888), 41.

Clam Chowder (Chowder de Lucines) (1894)

Prepare a quarter of a pound of well chopped fat pork, a small bunch of parsley chopped not too fine, four ounces of chopped onions, one and a half quarts of potatoes cut in seven-sixteenth of an inch squares; two quarts of clams retaining all the juice possible; one quart of tomatoes peeled, pressed and cut in half inch squares. Put the fat pork into a sauce pan, and when fried, add the onions to fry for one minute, then the potatoes, the clams and the tomatoes; should there not be sufficient moistening, pour in a little water and boil the whole until the potatoes are well done. Add five pilot crackers broken up into very small bits; one soup spoonful of thyme leaves, two ounces of butter, a very little pepper and salt to taste. This quantity will make four gallons, sufficient for sixty persons.

Another way.—Chop up a quarter of a pound of fat pork, melt it down, adding four ounces of onions cut in quarter inch dice, and fry them with pork, without coloring, then add one and a half pounds of potatoes cut in half inch squares, a pound of peeled and halved tomatoes, pressed out and cut in five-eighth inch squares, one ounce of coarsely chopped parsley, seventy-five medium sized clams, removing the hard parts and chopping them up very fine, the clam juice, a little salt if found necessary, pepper and thyme leaves. Boil the whole till the potatoes are cooked (the green part of celery chopped fine) and should the clam juice not be sufficient, then add a little water.

Source: Charles Ranhofer, *The Epicurean* (New York: R. Ranhofer, 1894), 267.

Clam Chowder—Coney Island Style (1893)

Take 1 quart of clams and their liquor—or a large can.
1 quart soup stock (or water).
1 quart raw potatoes cut in pieces.
1 large onion.
Butter size of egg.
A slice of ham—or knuckle bone.
1 pint tomatoes chopped.
1 teaspoon mixed thyme and savory.

6 cloves, 1 bay leaf, parsley.
1 teaspoon each black pepper and salt.

The different articles should be made ready and placed conveniently for use. Have the clams scalded and then cut in pieces and the liquor saved. Cut the potatoes in large squares and slice the onions. An hour before dinner put the butter and ham in a saucepan together, and the onions on top and set over the fire. Put the cloves inside of a little bunch of parsley and tie it on top of the onions, and, also the powdered or minced thyme and savory, and put on the lid, and let stew slowly. In about 15 or 20 minutes or before the ham and onions begin to brown put into the same saucepan the quart of soup stock, the clam liquor and potatoes, tomatoes, pepper and salt and let cook until the potatoes are done, then put in the cut clams. Take out the soup bunch and piece of ham, let boil up once with the clams in.

It is expected that the potatoes will sufficiently thicken this chowder without the use of flour but they should not be allowed to boil so much as to disappear altogether.

Source: Jessup Whitehead, *Cooking for Profit: A New American Cook Book,* 3rd ed. (Chicago: Jessup Whitehead, 1893), 98–99.

A Fulton Market Clam Chowder (1902)

Fry one-half pound of fat salt pork until the fat is extracted. Skim out the scraps and put in an onion chopped fine and fry it a light yellow. Turn the contents of the frying pan into a pot. Add a cupful of the strained liquor of the clams and the same quantity of water, which should have been heated together to the boiling point. Put in one quart of tomatoes stewed, the hard portions chopped. Put one-half dozen whole allspice and the same number of whole cloves into a piece of cheesecloth, tie securely, and drop into the soup. Cook four hours. Half an hour before serving add the hard part of the clams, which should have been chopped very fine, and ten minutes before the soup is to be put into the tureen add the soft part of the clams. Season with a dash of red pepper and a teaspoonful of Worcestershire sauce. This recipe is for fifty clams.

Source: Margaret Compton, *Grand Union Cook Book* (New York: Grand Union Tea Co., 1902), 51–52.

Fish Chowder (1902)

1 pound of fish
3 medium sized potatoes
1 pint of stewed tomatoes

1 tablespoonful of powdered thyme
1 pint of milk
1 pint of water
1 large onion
1 saltspoonful of celery seed
1 teaspoonful of salt
1 saltspoonful of pepper

Wash the fish; cut it into squares of one inch. Pare the potatoes and cut them into dice. Chop the onion. Put in the bottom of the kettle a layer of potatoes, then a layer of fish, then tomatoes, a sprinkling of onion, thyme, salt, pepper and celery seed, and so continue until the materials are used, having the last layer potatoes; add the water. Cover the kettle closely, and cook, without stirring, over a moderate fire for twenty minutes. In the meantime, heat the milk in a double boiler; add it quickly and serve. The tomatoes may be omitted.

Source: Sarah Tyson Rorer, *Mrs. Rorer's New Cook Book* (Philadelphia: Arnold, 1902), 93.

Coney Island Clam Chowder (1903)

Prepare salt pork the same as for Boston Clam Chowder. When fried enough put into a saucepan. Put in also a layer of thin raw sliced potatoes, a layer of raw clams, a very thin layer of canned tomatoes and a layer of broken crackers, a little pepper and salt. Repeat until the saucepan is about half full. Put in a very little white stock. Cover up tight. Let simmer on the side of the range one hour (be careful that it don't stick to the bottom); at the end of that time take off the cover and add more stock, enough to bring to the right consistency. Flavor with a little tomato catsup and Worcestershire. Serve.

Source: Joseph Vachon, *Vachon's Book of Economical Soups and Entreés* (Chicago: Hotel Monthly Press, 1903), 63–64.

Rhode Island Clam Chowder (1907)

Take half a peck of clams, and reserve the juice. Chop the hard part fine, and keep the soft parts whole. Cook the chopped part until tender in enough water to cover. Peel and slice two onions, also six potatoes, and add the chopped clams with two cupfuls of tomatoes. Add also the soft parts of the clams. Split a pound of Boston crackers, and arrange with the clams in layers in another kettle, seasoning each layer with pepper and powdered sweet herbs. Cover with cold water, and cook slowly until the vegetables are done.

Source: Olive Green [Myrtle Reed], *One Thousand Simple Soups* (New York: G. Putnam's Sons, 1907), 254.

Soft Clam Chowder (1907)

Tie in a muslin bag six whole allspice, six cloves, and six pepper-corns. Fry brown, with a sliced onion, one quarter of a pound of minced salt pork. Add six sliced potatoes, a can of tomatoes, the bag of spices, a pinch of red pepper, and four cupfuls of cold water. Simmer for four hours. Add a quart of soft clams parboiled and chopped fine, five Boston crackers that have been split and soaked in milk, simmer for five minutes, and serve very hot.

Source: Olive Green [Myrtle Reed], *One Thousand Simple Soups* (New York: G. Putnam's Sons, 1907), 255.

Connecticut Chowder (1907)

Chop fine a quarter of a pound of salt pork, and fry brown with two cupfuls of chopped onion. Add the juice drained from a quart of clams, two cupfuls of water, two cupfuls of canned tomatoes, four cupfuls of sliced potatoes, a cupful of fresh celery, cut fine, and a tablespoonful of powdered sweet herbs. Cook for thirty minutes, add pepper and salt to season, and a quart of chopped clams. Cook slowly for fifteen minutes.

Source: Olive Green [Myrtle Reed], *One Thousand Simple Soups* (New York: G. Putnam's Sons, 1907), 257.

Baked Chowder (1907)

Fry an onion in butter, with two pounds of fresh codfish cut into cubes. Butter a baking-dish and put a layer of fish into the bottom. Cover with a layer of raw potatoes, add a layer of minced salt pork, sprinkle with flour and minced parsley, and dot with butter. Repeat until the dish is full and add one cupful of tomato-juice. Add cold water to cover, and put on top a layer of split Boston crackers, that have been soaked in milk. Cover the dish, and bake for an hour, then take off the cover and brown.

Source: Olive Green [Myrtle Reed], *One Thousand Simple Soups* (New York: G. Putnam's Sons, 1907), 259.

Creole Corn Chowder (1907)

Fry brown in butter four large onions. Add five tomatoes, four sweet green peppers shredded, and two cupfuls of corn cut from the cob. Add boiling water to cover, season with salt, pepper, and sugar, and cook until the vegetables are done.

Source: Olive Green [Myrtle Reed], *One Thousand Simple Soups* (New York: G. Putnam's Sons, 1907), 260.

Bar Harbor Clam Chowder (1907)

Boil together for forty minutes one hundred soft clams, half a teaspoonful each of marjoram, thyme, sage, summer savory, and salt, a small onion, cut fine, five Boston crackers, split, half a can of tomatoes, and one third of a cupful of butter. Season with pepper, cloves, and curry powder, add one cupful of cream, and half a cupful of sherry.

Source: Olive Green [Myrtle Reed], *One Thousand Simple Soups* (New York: G. Putnam's Sons, 1907), 260.

Tomato Chowder Virginienne (1913)

Shred fine two ounces of salt pork, one white onion and one peeled green pepper. Saute together for about ten minutes, then add two ounces fresh butter and saute five minutes longer, but do not let get brown. Add two tablespoonfuls of flour, mix together, then add one pint of tomato sauce and one pint of white broth, three raw potatoes cut in small dice and twelve young tender fresh okras cut into half inch pieces. Let cook with a bouquet of garnishings for one hour. Add two ripe peeled fresh tomatoes and two dozen small raw oysters with the eyes cut off and a pinch of paprika, pepper and salt to taste and let boil five minutes longer.

Source: Archie Croydon Hoff, *Soups and Consommés* (Los Angeles: International Publishing, 1913), 47.

Clam Chowder (1915)

Mrs. Henry A. Young.

12 clams.
1 quart water.
4 potatoes, cut in dice.
1 can tomatoes.
1 carrot cut in dice.
6 slices bacon.
Stalk of celery, chopped.
1 tablespoon thyme.
1 onion chopped. Sprig parsley.
Salt and pepper.

Fry bacon and cut into dice; boil together one hour, liquor of clams, tomatoes, onions, selery, carrot, bacon, thyme and water; then add potatoes, clams (chopped), salt and pepper and boil half an hour. Add parsley just before serving. A cup of sherry adds greatly to the flavor.

Source: Women's Guild of Saint Paul's Church, *Tried and True Recipes* (Yonkers, New York: Herald Job Print, 1915), 46.

GAZPACHO

Gaspacha—Spanish (1824)

Put some soft biscuit or toasted bread in the bottom of a salad bowl, put in a layer of sliced tomatas with the skin taken off, and one of sliced cucumbers, sprinkled with pepper, salt, and chopped onions; do this until the bowl is full, stew some tomatas quite soft, strain the juice, mix in some mustard and oil, and pour over it; make it two hours before it is eaten.

Source: Mary Randolph, *The Virginia House-wife* (Washington: Davis and Force, 1824), 107.

Gazpacho (1845)

Chopped tomatoes and parsley; crackers and slices toasted bread; arrange each item in layers and between them place a small amount of minced mint leaves, powder sugar, good vinegar, a pinch of ground pepper. Cook to taste and serve cold.

Source: *Novisimo arte de cocina* (Philadelphia: Estereotipado é impreso por Compania, 1845), as translated by Mary Lou Dantona and Gloria Stiens, 72.

Gaspacho (1845)

Brown slices of bread and dip in vinegar, sugar and lemon; layer bread, chopped parsley; bread and fresh tomato; bread and olives and tornachiles; then a layer of bread and finally a layer of chicken breast. Cook until desired consistency.

Source: *Novisimo arte de cocina* (Philadelphia: Estereotipado é impreso por Compania, 1845), as translated by Mary Lou Dantona and Gloria Stiens, 221.

Gaspacho (1846)

Take two onions, some tomatas, a handful of green pimento, a cucumber, a clove of garlic, parsley and chervil; cut the whole into small pieces, and put them into a salad-bowl. Add as much crumbed bread as will form double the quantity which the dish already contains; season with salt, pepper, oil, and vinegar, like a salad, and complete the *gaspacho* with a pint of water to make the *bouillon*. *Gaspacho* is eaten with a spoon; it is a kind of raw soup. It is a favourite dish with the Andalusians, and is very refreshing and wholesome in their warm climate.

Source: [Louis Eustache Audot], *French Domestic Cookery* (New York: Harper & Brothers, 1846), 272–273.

Tomato Soup a L'andalouse (1896)

In a saucepan put one tablespoonful of butter, one-half of a carrot sliced, two tablespoonfuls of diced raw ham, one small bay leaf, a pinch of powdered thyme, six peppercorns, two cloves, one sprig of parsley and one stalk of celery. Cover and cook slowly for fifteen minutes; add two tablespoonfuls of flour and stir until well browned. Turn in one quart of fresh tomatoes, or one can, and stir until it boils. Add one scant teaspoonful of salt, one-quarter of a teaspoonful of pepper and one-half of a teaspoonful of sugar. Boil gently for three-quarters of an hour, and rub through a sieve. Return to the fire and add one teaspoonful of cornstarch dissolved in a little water. When thickened, add one pint of strong consommé, simmer for five minutes and serve.

Source: *Table Talk* 11 (November 1896):397.

GUMBO AND OKRA SOUPS

Soup (No Date)

Cut up Ocra, cymliss & Irish potatoes very small. Put them on in an earthen pot with water enough, some slices of lean bacon, onion and parsley chopped small, put in lima beans, & tomatas peel'd, boil a handful of thyme in it for about an hour, add a chicken & thicken the soup with flour & butter.

Source: Martha Randolph Jefferson, cookery manuscript, n.p., University of Virginia Accession no. 5385-4.

Okra Soup (1813)

One pound of Beef—1 doz Ochre—1 doz Tomatoes—one Onion, half a green pepper—early in the morning, put the Beef & Ochre in 3 gallons of water—and boil it to 3 pints—Skim it well add the Tomatoes &c—about 1 Oclock, for Diner at 3.

Source: Manuscript cookbook, Mrs. George Read, New Castle, 1813, n.p., in the Holcomb Collection, Historical Society of Delaware, Wilmington.

Gumbo Soup (1838)

Take four pounds of the lean of a fresh round of beef and cut the meat into small pieces, avoiding carefully all the fat. Season the meat with a little pepper and salt, and put it on to boil with three quarts and a pint of water (not more.) Boil it slowly and skim it well. When no more scum rises, put in half a peck of ochras, peeled and sliced, and half a peck of tomatas cut in quarters. Boil it slowly till the ochras and tomatas are entirely dissolved, and the meat all to rags. Then strain it through a cullender, and send it to table with slices of dry toast. This soup cannot be made in less than seven or eight hours. If you dine at two, you must put on the meat to boil at six or seven in the morning. It should be as thick as jelly.

Source: Eliza Leslie, *Directions for Cooking in Its Various Branches* (Philadelphia: Henry Carey Baird, 1838) 432–433.

Gumbo (1838)

Take an equal quantity of young tender ochras, and of ripe tomatas, (for instance, a quarter of a peck of each). Chop the ochras fine, and scald and peel the tomatas. Put them into a stew-pan without any water. Add a lump of butter, and a very little salt and pepper; and, if you choose, an onion minced fine. Let it stew steadily for an hour. Then strain it, and send it to table as soup in a tureen. It should be like a jelly, and is a favourite New Orleans dish. Eat dry toast with it. This gumbo is for fast days.

Source: Eliza Leslie, *Directions for Cooking in Its Various Branches* (Philadelphia: Henry Carey Baird, 1838), 439.

Gumbo (1841)

Peel two quarts of ripe tomatoes, mix with them two quarts of young pods of ochra, and chop them small; put them into a stew-pan, without any water; add four ounces of butter, and salt and pepper to your taste, and boil them gently and steadily for one hour; then pass it through a sieve into a tureen, and send to table with it, crackers, toasts, or light bread.

Source: Lettice Bryan, *The Kentucky Housewife* (Cincinnati: Stereotyped by Shepard & Stearns, 1841), 25.

Ochra Soup (1841)

Make a plentiful broth in the usual manner, of fresh beef, veal or poultry. Put into it equal proportions of ripe tomatoes and young ocras, having sliced the ocras very thin, and pared and sliced the tomatoes. Boil them gently till completely dissolved, pass it through a sieve and return it again to the pan. Have enough of the tomatoes and ocra to make it tolerably thick, season it to your taste with salt, cayenne and butter; and as soon as it comes to a boil, pour it into a tureen, on some small bits of toasted bread.

Source: Lettice Bryan, *The Kentucky Housewife* (Cincinnati: Stereotyped by Shepard & Stearns, 1841), 22–23.

Gumbo and Tomato Soup (1876)

If canned gumbo and tomatoes mixed are used, merely add to them a pint or more of stock or strong beef broth. Bring them to the boiling-point, and season with pepper and salt.

If the fresh vegetables are used, boil the tomatoes and gumbo together for about half an hour, first frying the gumbo in a little hot lard. Many, however, boil the gumbo without frying.

Source: Mary Henderson, *Practical Cooking and Dinner Giving* (New York: Harper & Brothers, 1876), 92.

Oyster Gumbo (1878)

Cut up a chicken, sprinkle with flour, and fry in the vessel in which the gumbo is made. When the chicken is nearly done, chop an onion and fry with it. Pour on this three quarts of boiling water and let it boil slowly till the flesh leaves the bones; then add the liquor from the oysters, salt and pepper to taste, two table-spoonfuls of tomato catchup; let this boil a short time, then add one hundred oysters, and allow them to boil only *five minutes.* When taken from the fire, and before pouring into the tureen, sprinkle in two table-spoonfuls of filé or sassafras powder.

Source: Ladies of the Saint Francis Street Methodist Episcopal Church, South, Mobile, Alabama, comp., *Gulf City Cook Book* (Dayton, Ohio: United Brethren Publishing House, 1878), 13.

Chicken Gumbo with Oysters (1902)

1 young chicken
1 good-sized tomato
A bay leaf
2 tablespoonfuls butter
1 quart of green okra
25 oysters
1 teaspoonful of salt
1 saltspoonful of pepper

Draw the chicken, and cut it up as for a fricassee. Put into the soup kettle the two tablespoonfuls of butter; when hot (not brown) throw in the chicken and shake until the chicken is browned. Draw to one side, and add the okra that has been washed and cut into slices. Cook slowly with the chicken and butter over a mild fire for ten or fifteen minutes; then add two quarts of boiling water, the tomato cut into slices, and the bay leaf; bring to boiling point, and simmer gently for one hour or until reduced to a quart and a pint; now add the oysters, salt and pepper; bring to boiling point, and serve with carefully boiled rice. The rice is usually passed, allowing each one to put a spoonful into the gumbo.

Source: Sarah Tyson Rorer, *Mrs. Rorer's New Cook Book* (Philadelphia: Arnold, 1902), 84.

Southern Gumbo (1916)

3/4 lb. veal, or 1 small chicken
½ lb. ham
1 can tomatoes
1 pint fresh oysters, or 1 can shrimps
1 large tablespoon crisco
1 Onion
1 quart water
1 lb. fresh Okra, or 1 can okra

Cut meat in small pieces and fry in crisco until nicely browned, add salt and pepper to taste, one sliced onion, one quart of water, one lb. of fresh okra, cook one hour. If canned okra is used, add after meat is thoroughly cooked, then can tomatoes. Just before serving, add one can shrimps or 1 pint fresh oysters, cook for six minutes and serve with boiled rice.—Mrs. R. H. Hawley.

Source: *The Garfield Woman's Club Cook Book* (Garfield, Utah: Garfield Woman's Club, 1916), 9–10.

RECIPES FOR OTHER SOUPS WITH TOMATOES AS AN INGREDIENT

Barley Soup (1824)

Put on three gills of barley, three quarts of water, a few onions cut up, six carrots, scraped and cut in dice, an equal quantity of turnips cut small: boil it gently two hours, then put four or five pounds of the rack or neck of mutton, a few slices of lean ham, with pepper and salt; boil it slowly, two hours longer, and serve it up. Tomatas are an excellent addition to this soup.

Source: Mary Randolph, *The Virginia House-wife* (Washington: Davis and Force, 1824), 33.

Tomato Mutton Soup (1839)

Procure a *rack* of good wedder mutton, put in a pot or stewpan, with four quarts of water, add six or eight tomatos, peeled and sliced, two onions chopt fine, two potatoes, and as many turnips and carrots, sliced; a small bunch of thyme and sweet basil; wrap up an onion in paper and roast it thoroughly in hot embers, add it one hour before the soup is done, cook for three or four hours; add plenty of salt and black pepper; skim off the fat.

Source: Edward James Hooper, *The Practical Farmer, Gardener and Housewife* (Cincinnati: Geo. Conclin, 1839), 495.

Corn and Tomato Soup (1873)

Take a good soup-bone; wash it nicely; pour over it sufficient water to cover it well; cut up an onion in it; salt and pepper; cut down about one dozen ears of corn and as many tomatoes in it, and let it boil slowly for at least three hours. For dumplings, take one egg and beat it a little, one coffeecup sour milk, small teaspoonful of soda, a little salt, and flour enough to make a stiff batter; drop it into the boiling soup, from a spoon, twenty minutes before serving. These dumplings are good in bean soup also.

Source: *Presbyterian Cook Book* (Dayton, Ohio: Crooke, 1873), 19–20.

Macaroni Soup (1873)

Take six pounds of beef, and put into four quarts of water, with two onions, one carrot, one turnip, and a head of celery; boil it down three or four hours slowly, till there is about two quarts of water; then let cool. Next day, half an hour before dinner, take off the grease and pour the soup into the kettle (leaving the sediment out), and add salt to suit the taste; a pint of macaroni broken into inch pieces, and a tablespoonful and a half of tomato catsup.

Source: *Presbyterian Cook Book* (Dayton, Ohio: Crooke, 1873), 20.

Sago and Tomato Soup (1875)

Boil two quarts of peeled, sliced tomatoes, and a sliced onion, until half cooked. Pour a pint and a half of boiling water on a gill of sago. Let it boil ten minutes, then put it with the tomatoes, and add a quart of boiling water; season with two tablespoonfuls of salt, three of sugar, a teaspoonful of pepper, and four cloves, and boil until the tomatoes are done; if too thick, add boiling water, and more seasoning if liked. When the soup is in the tureen, strew it with bread dice.

Source: [Elizabeth Smith Miller], *In the Kitchen* (Boston: Lee and Shepard; New York: Lee, Shepard, and Dillingham, 1875), 49–50.

Rice and Tomato Soup (1877)

Strain, and pass through a sieve with a wooden spoon, one pint of tomatoes, either fresh or canned, stir them into two quarts of good, clear stock, free from fat; season it with a teaspoonful of salt, and quarter of a saltspoonful of pepper; taste, and if seasoning seems deficient add a little more, but do not put in too much for general liking, for more can easily be added, but none can be taken out. Add four ounces of rice, well washed in plenty of cold water, and boil the soup slowly for three quarters of an hour before serving.

Source: Juliet Corson, *Miss Corson's Cooking Manual of Practical Directions for Economical Every-Day Cooking* (New York: Dodd, Mead, 1877), 26.

Cubion (1878)

Stir into two table-spoonfuls of hot lard sufficient flour to brown it, and red and black pepper to taste. Then add eight onions, sliced, a large bunch of parsley chopped fine, a little thyme, and one quart of tomatoes. Let these cook fifteen minutes, stirring all the time; then add two quarts of boiling water, and boil slowly for three hours. Three quarters of an hour before serving add one quart of claret, one table-spoonful of whole allspice, and one large lemon, sliced thin. Cut your red-fish in pieces about three or four inches square, bones and all, and half an hour before serving put it in and let it boil until dinner. Serve with rice, as gumbo. Salt to taste.

Source: Ladies of the Saint Francis Street Methodist Episcopal Church, South, Mobile, Alabama, comp., *Gulf City Cook Book* (Dayton, Ohio: United Brethren Publishing House, 1878), 11–12.

Green-Corn-and-Tomato Soup (1880)

This requires two pounds of the neck of beef, a quart of sliced tomatoes, a quart of corn sliced from the cob, three pints of water, one tablespoonful of butter, one of flour, and salt and pepper to suit the taste. Put the meat and water into a soup-pot, and as soon as the liquor begins to boil, skim it carefully. Simmer for three hours; then add the tomato and the corn-cobs. Cook for half an hour; then strain into another kettle, and add the corn, the flour and butter mixed together, and enough salt and pepper to season well. Cook forty minutes longer.

Source: Maria Parloa, *Miss Parloa's Kitchen Companion* (Boston: Dana Estes, 1880), 141.

Macaroni-and-Tomato Soup (1880)

The materials required are two pounds of the neck of beef, three quarts of water, one pint of stewed tomato, one pint of macaroni broken into two-inch pieces, an onion, two cloves, a sprig of parsley, half a cupful of cornstarch, two table-spoonfuls of salt, and half a teaspoonful of pepper.

Have the meat perfectly clean and broken into small pieces. Put it into a soup-pot with the cold water, and heat it slowly to the boiling-point; then skim carefully, and simmer for two hours. At the end of that time add the onion, parsley, and clove, and cook for an hour longer. After skimming off all fat from the soup, mix the cornstarch with a cupful of cold water, and stir into the soup. Now add the tomato, salt, and pepper, and cook gently for half an hour longer. Wash the macaroni in cold water, and put it into a stew-pan with a quart of boiling water. Boil rapidly for twenty minutes; then turn the macaroni into a colander, and pour a quart of cold water over it. Strain the soup, and return it to the kettle; then add the macaroni, and cook for twenty minutes. Serve toasted bread with the soup.

Source: Maria Parloa, *Miss Parloa's Kitchen Companion* (Boston: Dana Estes, 1880), 142.

Sago and Tomato Soup (1884)

Boil two quarts of beef-broth in a saucepan; stir, and let drop into it, rain-like, four ounces of sago; boil twenty minutes, add a quart of tomato purée (No. 197). Boil five minutes longer, skim, pour in a soup-tureen, and serve.

Source: Felix Déliée, *The Franco-American Cookery Book* (New York: G. P. Putnam's Sons, 1884), 391.

Tapioca and Tomato Soup (1884)

Boil two quarts of beef-broth in a saucepan; let drop like rain four ounces of pulverized tapioca, stirring with a dressing-spoon in the right hand; cook slowly for twenty minutes, skim, add a quart of tomato purée, boil a little longer, skim again, pour into a soup-tureen, and serve.

Source: Felix Déliée, *The Franco-American Cookery Book* (New York: G. P. Putnam's Sons, 1884), 540.

Chicken and Tomato Soup (1884)

Take a large, tender chicken; singe, draw, and cut in small pieces; put in a saucepan with two ounces of butter, two ounces of lean ham, and an onion cut small; stir on the fire until light brown, strain off the butter, wet with two quarts of beef-broth and a pint of water; add salt, white and a pinch of red pepper, and a bunch of parsley with aromatics; scald, peel, press the seeds out, and cut a dozen tomatoes in quarters; put them with the soup, cover, and cook an hour; skim the fat, remove the parsley, pour in a soup-tureen, and serve.

Source: Felix Déliée, *The Franco-American Cookery Book* (New York: G. P. Putnam's Sons, 1884), 434.

Peanut Tomato Soup (1899)

Boil the peanuts as directed. Sift through a colander to make them perfectly smooth. To 1 cup of sifted peanuts add ½ cup of sifted tomatoes, ½ cup of water, and salt to taste. Thicken with 1 teaspoonful of white flour to keep it from settling. If it is too thick, thin with water.[2]

Source: Almeda Lambert, *Guide for Nut Cookery* (Battle Creek, Michigan: Joseph Lambert, 1899), 304.

Tomato Nutmeatose Soup (1899)

Take one rather small onion and grate fine. Add 1 cup of sifted tomatoes, 1 cup of grated nutmeatose, 1 cup of water, and salt to taste. Thicken with a very little gluten.[3]

Source: Almeda Lambert. *Guide for Nut Cookery* (Battle Creek, Michigan: Joseph Lambert, 1899), 304.

Tomato Celery Soup (1899)

Take one rather small onion grated, 2 tomatoes, 1 celery head, 1 carrot sliced, 1 spoonful of brown sugar, and salt to taste. Cook until tender and serve with croutons.

Source: Almeda Lambert, *Guide for Nut Cookery* (Battle Creek, Michigan: Joseph Lambert, 1899), 305.

Vermicelli Soup with Tomato Purée (1896)

Prepare three quarts of fish stock, as for thick soup; when boiling move it to the side of the fire and let it simmer for half an hour. Make one and one-half pints of fresh tomato purée. Skim the fat off the soup, put in a bunch of parsley and sweet herbs and the tomato purée, then allow it to simmer for twenty minutes longer. Boil gently in salted water one-half pound of vermicelli. Strain the soup, put in the vermicelli, skim off all the fat and boil up again. Turn the soup into a tureen, and serve.

Source: Oscar Tschirky, *The Cook Book by "Oscar" of the Waldorf* (Chicago and New York: Werner, 1896), 45.

Albóndigas a la Español (1898)
Spanish meat balls

Chop pork loin until it is minced or ground, and remove any sinews. Then add green onion, peeled and seeded tomatoes, parsley, fresh or dry cilantro, and garlic, all finely chopped. To this add a piece of cornmeal paste (nixtamal), or a spoonful of cornmeal, one or two raw eggs, bread crumbs, a piece of lard or butter, salt and pepper. Mix all these together by hand, forming meatballs as usual, then add them to the stock, which should be boiling. If there is no stock, cook in boiling water. Season the broth with tomato, green onion, chopped parsley, salt, pepper and butter. Thicken the broth with beaten egg yolks when ready to serve.

Source: Encarnación Pinedo, *The Spanish Cook: A Selection of Recipes from Encarnación Pinedo's* El cocinero español, *Edited and Translated by Dan Strehl* (Pasadena: Weather Bird Press, 1992), 2.

Cordero [Lamb Soup] (1906)

Cut a pound of young lamb into small chunks and fry with a sliced onion in hot lard. When nicely browned add three peeled and sliced tomatoes and three green peppers chopped fine. Cover with two quarts of water and simmer slowly; add a cupful of green peas, one of green corn cut from the cob, a half-cupful of rice, salt and chile pepper. Work into a raw egg a teaspoonful of oil and a half-teaspoonful of vinegar; put this in the bottom of the soup-tureen and pour the soup over it.

Source: May E. Southworth, comp., *One Hundred and One Mexican Dishes* (San Francisco: Paul Elder, 1906), 4.

Barley and Tomato Soup (1908)

Cook 1 can of tomatoes and 1 chopped Spanish onion together for fifteen minutes, then rub through a wire sieve; add 3 tablespoons of pearl barley, 1 tablespoon of butter, some pepper and salt and cook for one hour, until the barley is soft. Re-season before serving.

Source: M.R.L. Sharpe, *The Golden Rule Book: Six Hundred Recipes for Meatless Dishes* (Cambridge: University Press, 1908), 48.

Spinach-Tomato Soup (1908)

Put 1 tablespoon of butter into the frying pan, and when melted add 1 onion chopped fine, and let cook slowly for ten minutes. Put 1 cup of cold prepared spinach into the butter and onion and 1 cup of tomato sauce or tomatoes, and let heat through. Put 2 cups of milk in a double boiler with 1 tablespoon of flour and 1 of butter rubbed together. Add a pinch of soda to the tomato-spinach mixture, press it through a sieve, and stir the purée into the milk when it boils. Season with salt and pepper and add 1 tablespoon of cream.

Source: M.R.L. Sharpe, *The Golden Rule Book: Six Hundred Recipes for Meatless Dishes* (Cambridge: University Press, 1908), 67.

Chicken-and-Tomato Bouillon (1917)

Cut half an onion and one-fourth a carrot in very thin slices, and sauté these in a little butter or dripping until yellowed and softened. Add part of a bay leaf, a sprig of parsley, and a bit of yellow lemon rind. Let simmer in a pint of water half an hour. Then add to one quart of chicken broth with all the liquid that can be drained from a can of tomatoes. Mix with these the crushed shells of several eggs and the slightly beaten whites of two, and salt and pepper as needed. Stir constantly over the fire until the boiling-point is reached. Let boil five minutes, then keep hot without boiling about ten minutes. Skim and strain through several folds of cheese-cloth laid over a wire strainer. Nothing but the liquid from the can of tomatoes is used; this is secured by draining them in a sieve without pressure.

Source: Janet Hill, "Seasonable and Tested Recipes," *American Cookery* 21 (January 1917):449–450.

NONSOUP RECIPES USING TOMATO SOUP AS AN INGREDIENT

Tomato Soup with Spaghetti (1906)

To one quart of beef or lamb broth add one pint of tomatoes, pressed through a sieve fine enough to keep back the seeds, a teaspoonful of sugar, and salt to season. Then put over the fire to boil. Have ready four ounces of spaghetti, boiled fifteen minutes in rapidly boiling, salted water, then drained and rinsed in cold water. Spread on a cloth, and cut in inch lengths. Put the spaghetti in the soup. Let cook ten minutes, then skim and serve. Pass with the soup grated Parmesan cheese, on a plate, that those who wish may add a spoonful to their plate of soup. Or cut milk rolls (yeast) in slices. Toast these, then spread one side with butter, and sprinkle over the butter Parmesan cheese and paprika. Set the slices into a hot oven to melt (not brown) the cheese, then pass at once on a plate covered with a napkin.

Source: Janet M. Hill, "Seasonable Recipes," *Boston Cooking-School Magazine* 10 (March 1906):382–383.

Tomato Soup on Toast (1915)

Take 1 can condensed tomato soup, 2 tablespoonsful flour, 2 tablespoonsful butter, 1½ cups milk, 8 slices of flour toast. Melt the butter and add the flour. Pour in the milk gradually. When smooth add the can of soup undiluted, when the mixture is hot, serve at once on the toast. A spicy and pleasing novelty.

Source: *The Eagle Cook Book* (Brooklyn: Brooklyn Evening Eagle, 1915), n.p.

Tomato Soup Jelly (1915)

To one can of condensed soup and one-half can of boiling water, one teaspoon of sharp vinegar, or sherry if preferred; one-half of a green pepper not too finely chopped, and bring all to the boiling point. Season to taste with paprika and salt. Dissolve one-half box of gelatin in cold water, and stir well. Pour into soup cups and when cool place in refrigeration until chilled. Serve very cold.

Source: *Des Moines Evening Capitol*, April 13, 1915.

Spanish Veal Balls (1916)

1½ pound chopped raw veal.
½ pound chopped salt pork.
½ cup fine bread crumbs (packed down).
1 rounded teaspoonful salt.
½ teaspoonful poultry seasoning.
½ cup finely chopped green pepper.
1 teaspoonful chopped parsley.
1 large onion, finely chopped.
1 egg, well beaten.
1 can CAMPBELL'S TOMATO SOUP.

Mix meat, seasonings, crumbs and egg. Shape into flat balls. Roll in flour, and brown thoroughly in saute pan with a little pork or other fat. Turn into shallow baking pan, and pour over CAMPBELL'S TOMATO SOUP, adding a little water if necessary, a few drops Worcestershire sauce, grated onion, and chopped parsley. Finish cooking slowly for one hour in oven. Serve on hot platter with parsley garnish on each ball. (Makes twelve large balls.)

Source: *Helps for the Hostess* (Camden, New Jersey: Joseph Campbell Company, ca. 1916), 18.

Bluefish with Tomato Sauce (1916)

3 pounds bluefish, bones removed.
½ teaspoonful each salt and curry powder.
½ can CAMPBELL'S TOMATO SOUP.
1 tablespoonful butter.
½ teaspoonful anchovy paste.
½ finely chopped onion.
½ teaspoonful chopped parsley.
Juice of half a lemon.

Butter a baking dish, lay in bluefish, season with salt, curry and pepper. Mix butter, paste, onion, parsley and lemon juice, add to tomato soup, and pour around fish. Bake about forty-five minutes, and serve in the same dish.

Source: *Helps for the Hostess* (Camden, New Jersey: Joseph Campbell Company, ca. 1916), 19.

Scallop Cocktail (1916)

1 pint scallops.
8 tablespoonfuls CAMPBELL'S TOMATO SOUP.
1 teaspoonful salt.
½ teaspoonful white pepper.
1 teaspoonful dry mustard.
1 teaspoonful olive oil.
1 teaspoonful each finely minced parsley and onions.
8 drops tabasco sauce.
1½ teaspoonfuls Worcestershire sauce.
1 teaspoonful grated horseradish.
2 teaspoonfuls vinegar.

Cook scallops five minutes, chill and cut in halves. Add to well-mixed sauce[;] serve in individual scallop shells on bed of cracked ice.

Source: *Helps for the Hostess* (Camden, New Jersey: Joseph Campbell Company, ca. 1916), 34.

Chilaly (1916)

1 tablespoonful butter.
½ teaspoonful dry mustard and salt.
1 teaspoonful Worcestershire sauce.
2 tablespoonfuls chopped green pepper.
1 tablespoonful chopped onion.
½ cup CAMPBELL'S TOMATO SOUP.
½ pound sharp cheese run through chopper.
2 tablespoonfuls milk.
1 egg.

Cook butter with onions and pepper until brown. Rub mustard and salt together, add to TOMATO SOUP, then add to butter mixture. Boil up, add cheese and Worcestershire. When cheese is melted, add milk, and egg slightly beaten. Serve at once on hot pilot crackers. This is best cooked in the blazer of a chafing dish.

Source: *Helps for the Hostess* (Camden, New Jersey: Joseph Campbell Company, ca. 1916), 38.

Green Corn, Creole Style (1916)

2 cups corn kernels cut from cob.
1 cup CAMPBELL'S TOMATO SOUP.
4 tablespoonfuls butter.
4 tablespoonfuls flour.
1 teaspoonful salt.
1 tablespoonful onion juice.
2 tablespoonfuls chopped green pepper.
1 tablespoonful chopped parsley.

Melt butter and in it cook the pepper until tender. Add the flour, blend, and add tomato soup. Stir until smooth. Then add seasonings and corn, beat thoroughly, and serve at once.

Source: *Helps for the Hostess* (Camden, New Jersey: Joseph Campbell Company, ca. 1916), 39.

Oriental Roast (1916)

1 cup lentils.
⅛ pound grated American cheese.
Salt, paprika, bread crumbs.
½ cup CAMPBELL'S TOMATO SOUP.
1 tablespoonful each butter and flour.
1 teaspoonful each chopped onion and green pepper.

Soak lentils over night. Change to fresh water, and cook lentils until soft in as little water as possible. Rub through fine strainer, add cheese, onion, pepper, seasoning, and enough bread crumbs to mould into a roll. Lay on buttered paper in baking dish, and bake until moderate brown, basting with melted butter. Melt one tablespoonful butter, rub in one tablespoonful flour, add tomato soup, and season. Pour over the lentil roll, serve very hot.

Source: *Helps for the Hostess* (Camden, New Jersey: Joseph Campbell Company, ca. 1916), 50.

Quick Tomato Bouillon (1924)

5 cupfuls boiling water
3 teaspoonfuls vegetable or meat extract or 3 bouillon cubes
⅔ cupful chili sauce or ½ can tomato soup

Put the extract, or bouillon cubes, whichever is used, into a saucepan; add the boiling water and simmer just long enough to dissolve. Add the sauce or soup, and bring again to boiling point, then strain through a fine sieve.

Source: Ida C. Bailey Allen, *Mrs. Allen on Cooking, Menus, Service* (Garden City, New York: Doubleday, Page, 1924), 129.

Italian Spaghetti (1928)

1 can Campbell's tomato soup
1 box spaghetti
¼ pound American cheese
½ pound bacon
3 onions

Dice the bacon and crisp, add the onions, cut fine. Add the soup. In the meantime, have spaghetti cooking, so that it will be done when the other mixture is ready. Twenty minutes boiling is sufficient. Pour cold water through it. Warm again. Pile on a platter. Pour the sauce over it, in which the cheese is melted.—H. M. Witchie.

Source: Minneapolis Public Library Staff Association, comp., *Library Ann's Cook Book* (Minneapolis, Minnesota: n.p., 1928), 43.

Ashville Salad (1928)

1 can Campbell's tomato soup
3 Philadelphia cream cheeses
2 tablespoons Knox unflavored gelatin
½ cup cold water
1 cup mayonnaise dressing
1½ cups chopped celery, peppers, olives, etc.

Bring soup to the boiling point, add cheese, and stir until smooth, then add gelatin which has stood for half an hour in the cold water. Stir well and let it cool before adding mayonnaise and vegetables.—Gladys Moriette Wilson, Irene Melgaard Hauser.

Source: Minneapolis Public Library Staff Association, comp., *Library Ann's Cook Book* (Minneapolis, Minnesota: n.p., 1928), 46–47.

Quick Tomato Spice Cake (1970)

1 package (2 layer) spice cake mix
1 can (10¾ ounces) condensed tomato soup
½ cup water
2 eggs

Mix *only* the above ingredients; following directions on package. If desired, fold in 1 cup chopped walnuts. Bake as directed. Frost with Cream Cheese Frosting or other favorite white dressing.

Source: *Easy Ways to Delicious Meals: 465 Quick-to-Fix Recipes Using Campbell's Convenience Foods*, rev. ed. (Camden, New Jersey: Campbell Soup Company, 1970), n.p. Used with permission.

Notes

1. PRIMORDIAL SOUP

1. Reay Tannahill, *Food in History*, 2nd ed. (New York: Crown Publishers, 1989), 14–15.
2. Tannahill, *Food in History*, 15–16; Herodotus, Book IV, sections 60–61, as in *The Histories* (London and New York: Penguin Books, 1996), 234.
3. E. W. Gifford, "The Southeastern Yavapai," *American Archaeology and Ethnology* 29 (1932):205.
4. Frances Densmore, "Chippewa Customs," Bulletin no. 86 (Washington, D.C.: Smithsonian Institution Bureau of American Ethnology, 1929), 41.
5. Paul LeJeune, "Relation of What Occurred in New France in the Year 1633," as in Reuben Gold Thwaites, ed., *The Jesuit Relations and Allied Documents: Travels and Explorations of the Jesuit Missionaries in New France, 1610–1791,* 73 vols. (Cleveland: Burrows Bros., 1896–1901), 5:97.
6. Pliny Earle Goddard, "Life and Culture of the Hupa," *American Archaeology and Ethnology* 1 (1903–1904):23; Maria-Louise Sideroff, letter to the author, July 17, 1998.
7. Fritz Ruf, ed., *"Die sehr bekannte dienliche Löffelspeise" Mus, Brei und Suppe—kulturgeschichtlich betrachtet* (Velbert-Neviges, Germany: BreRing Verlag, 1989), 82, as translated by Michael Beiser.
8. Germaine Tillion [Quintin Hoare, trans.], *The Republic of Cousins* (London: Al Saqi Books, 1983), 79.
9. C. Anne Wilson, "Pottage and Soup as Nourishing Liquids," in C. Anne Wilson, ed., *Fifth Leeds Symposium on Food and Society: Liquid Nourishment* (Edinburgh: Edinburgh University Press, 1993), 4.
10. Andrew Dalby, *Siren Feasts: A History of Food and Gastronomy in Greece* (London and New York: Routledge, 1996), 22, 90.
11. John Edwards, *The Roman Cookery of Apicius: A Treasury of Gourmet Recipes and Herbal Cookery* (Point Roberts, Washington: Hartly & Marks, 1984), 94–95; Ilaria Gozzini Giacosa [Anna Herklotz, trans.], *A Taste of Ancient Rome* (Chicago and London: University of Chicago Press, 1992), 80–84.
12. Giacosa, *A Taste of Ancient Rome,* ix.
13. Dalby, *Siren Feasts,* 201–203; Günay Kut, "Turkish Culinary Culture," in Semahat Arsel, ed., *Timeless Tastes: Turkish Culinary Culture,* 2nd ed. (Istanbul: Vehbi Koç Kakh, 1996), 41; Nezihe Araz, "Eat Sweet, Talk Sweet," ibid., 28; Tuğrul Şavkay, "The Cultural and Historic Context of Turkish Cuisine," ibid., 84.
14. John Ayto, *Food and Drink from A to Z: A Gourmet's Guide,* 2nd ed. (Oxford and New York: Oxford University Press, 1993), 44.
15. Terence Scully, *The Art of Cookery in the Middle Ages* (Woodbridge, U.K.: Boydel Press, 1995), 170.
16. Robert E. Lewis, "Middle English Culinary Terms, II. Soupe," Culinary Historians of Ann Arbor *Newsletter* 4 (Summer 1990):2–3, 5.

17. Odile Redon, François Sabban, and Silvano Serventi, *The Medieval Kitchen: Recipes from France and Italy* (Chicago and London: University of Chicago Press, 1998), 207–208.
18. James Cross Giblin, *From Hand to Mouth; or, How We Invented Knives, Forks, Spoons, and Chopsticks and the Table Manners to Go with Them* (New York: HarperCollins, 1987), 28–29.
19. T. Sarah Peterson, *Acquired Taste: The French Origins of Modern Cooking* (Ithaca and London: Cornell University Press, 1994).
20. Mary Ella Milham, ed., *Platina: On Right Pleasure and Good Health* (Tempe, Arizona: Medieval & Renaissance Texts & Studies, 1998), 1–59, 295, 323, 337, 339–341, 343.
21. Bartolomeo Scappi, *Opera dell'arte del cucinare*, reprint (n.p.: Arnaldo Forni Editore, 1981); Ayto, *Food and Drink from A to Z*, 220; John Mariani, *The Dictionary of Italian Food and Drink* (New York: Broadway Books, 1998), 154.
22. Scully, *The Art of Cookery in the Middle Ages*, 128–129.
23. Vance Packard, *Hidden Persuaders* (New York: David McKay, [1957]), 102.
24. Scully, *The Art of Cookery in the Middle Ages*, 126–127.
25. *The Forme of Cury*, as in Richard Warner, *Antiquitates Culinariæ; or, Curious Tracts Relating to the Culinary Affairs of the Old English* (London: R. Blamine, 1791), 24.
26. Thomas Austin, ed., *Two Fifteenth-Century Cookery-Books* (London: Oxford University Press, 1996), 11, 90.
27. *Satyr against the French* (1691), 16, as noted in the *Oxford English Dictionary.*
28. John Gay, as quoted in Samuel Johnson, *A Dictionary of the English Language: In Which the Words Are Deduced from Their Originals, and Illustrated in Their Different Significations by Examples from the Best Writers,* 2nd ed., 2 vols. (London: Printed by W. Strahan, 1773), 2: n.p.
29. Jonathan Swift, "A Panygyrick on the Dean," in *The Works of Dr. Jonathan Swift*, 22 vols. (London: C. Davis, C. Hitch, L. Hans et al., 1754), 7:197.
30. Andrew Broode, as quoted in J. C. Drummond and Anne Wilbraham, *The Englishman's Food* (London: Readers Union/Jonathan Cape, 1958), 51.
31. Gervase Markham, *The English Hus-wife* (London: John Beale, 1615), reprint Michael R. Best, ed., *The English Housewife* (Kingston and Montreal: McGill-Queen's University Press, 1986), 65, 74, 91–92.
32. *The Closet of the Eminently Learned Sir Kenelme Digby, K.T., Opened* (London: Printed by E. C. for H. Brome, 1669), 141–158, 166–167, 169–170, 172–174.
33. Matthew Hamlyn, ed., *The Recipes of Hannah Woolley: English Cookery of the Seventeenth Century* (London: Heinemann Kingswood, 1988), 102–103.
34. Robert May, *The Accomplished Cook* (London: R. W., 1660), 13–15, 47–48, 52, 65, 67, 77–78, 85, 93–94, 102, 109, 303–304, 426, 436, 450, 454–455.

35. E. Smith, *The Compleat Housewife; or, Accomplished Gentlewoman's Companion*, 15th ed. (London: R. Ware, etc., 1753), 26–31, 86–94.
36. Charles Carter, *The Complete Practical Cook; or, A New System of the Whole Art and Mystery of Cookery* (London: W. Meadows, 1730), 32.
37. Hannah Glasse, *The Art of Cookery Made Plain and Easy* (London, 1747), facsimile (London: Prospect Books, 1983), 62–68, 77–78, 90, 120–121, 123.
38. Glasse, *The Art of Cookery Made Plain and Easy*, 65; Ayto, *Food and Drink from A to Z*, 324; Johnson, *A Dictionary of the English Language*, 2: n.p.
39. Hannah Glasse, *The Art of Cookery Made Plain and Easy*, 8th ed. (London: Printed for A. Millar and J. and R. Tonson, 1763), 262.
40. Richard J. Hooker, *The Book of Chowder* (Boston: Harvard Common Press, 1978), 9, 27; Karen Hess, "Historical Glossary," in Mary Randolph, *The Virginia Housewife with Historical Notes and Commentaries by Karen Hess*, facsimile (Columbia: University of South Carolina Press, 1984), 265–267.
41. Elizabeth Raffald, *The Experienced English Housewife* (Manchester: Printed by J. Harrop for the author, 1769), 1–11; J. C. Drummond and W. R. Lewis, "The Examination of Some Tinned Foods of Historic Interest," *Chemistry and Industry*, (August 27, 1938):809.
42. A[lexander] Hunter, *Culina Famulatrix Medicinae; or, Receipts in Modern Cookery, with a Medical Commentary*, 4th ed. (York: T. Wilson and R. Spence, 1806), 16–17, 19–22, 24–25, 31–36, 40–43, 45–46, 57–58, 63–68, 71, 78–79, 87, 92–94, 98–102, 104–106, 116, 125–127, 138–141, 160–175, 182–184, 194, 198, 219, 227–228, 232–233, 245, 247, 250–251, 271–272, 276–277.
43. Richard Dolby, *The Cook's Dictionary and Housekeeper's Directory* (London: Henry Colburn and Richard Bentley, 1833), 88–89, 191, 192, 277, 337, 440–441, 480–481, 510.
44. Giles MacDonogh, *A Palate in Revolution: Grimod de La Reynière and the* Almanach des Gourmands (London and New York: Robin Clark, 1987), 50; Ruf, ed., *"Die sehr bekannte dienliche Löffelspeise,"* 62, as translated by Michael Beiser.
45. Ruf, ed., *"Die sehr bekannte dienliche Löffelspeise,"* 58, as translated by Michael Beiser.
46. Benjamin Delessert, *Sur les fourneaux à la Rumford et les soupes économiques* (Paris: Laoli, 1800).
47. Ruf, ed., *"Die sehr bekannte dienliche Löffelspeise,"* 60, as translated by Michael Beiser.
48. Drummond and Wilbraham, *The Englishman's Food*, 257–258.
49. Helen Morris, *Portrait of a Chef: The Life of Alexis Soyer* (Cambridge: Oxford University Press, 1938), 74–82; Elizabeth Ray, *Alexis Soyer: Cook Extraordinary* (Lewes, South Sussex: Southover Press, 1990), 43–49.
50. F. Volant and J. R. Warren, comps. and eds., *Memoirs of Alexis Soyer* (London: W. Kent, 1859), 102–107. Unbeknownst to Soyer or others at the time, his soup recipes may well have made a bad situation worse. Recent analysis by Joseph Carlin indicated that his recipes were deficient in the amino acid lysine.

51. Morris, *Portrait of a Chef: The Life of Alexis Soyer*, 79–80.
52. Charles Elmé Francatelli, *A Plain Cookery Book for the Working Classes* (London: Routledge, Warne, and Routledge, 1852), 63.
53. A. Soyer, *Shilling Cookery for the People* (London: Geo. Routledge, 1855), 7.
54. Richard J. Hooker, *A History of Food and Drink in America* (Indianapolis and New York: Bobbs-Merrill, 1981), 79–80.
55. Karl Theodor Eben, *Gottlieb Mittelberger's Journey* (Philadelphia: J. J. McVey, 1898), 65.
56. Johannes Schweizer in Robert H. Billigmeier and Fred Altschuker Picard, eds., *The Old Land and the New: The Journals of Two Swiss Families in America in the 1820s* (Minneapolis: University of Minnesota Press, 1965), 76.
57. George Towle, *American Society*, 2 vols. (London: Chapman and Hall, 1870), 1:270.
58. *Gentleman's Magazine*, as quoted in Helen Bullock, *The Williamsburg Art of Cookery* (Williamsburg: Colonial Williamsburg, 1979), 7–8.
59. William Byrd, *Histories of the Dividing Line betwixt Virginia and North Carolina*, with introduction and notes by William K. Boyd (Raleigh: North Carolina Historical Commission, 1929), 163, 187, 225, 254, 287.
60. Raffald, *The Experienced English Housewife*, 1–11; Drummond and Lewis, "The Examination of Some Tinned Foods of Historic Interest," 809.
61. E. Smith, *The Compleat Housewife* (Williamsburg: William Parks, 1742), 141.
62. Susannah Carter, *The Frugal Housewife* (Philadelphia: James Carey, 1796), 66–73.
63. Anne Gibbons Gardiner, ed., *Mrs. Gardiner's Family Receipts from 1763* (Boston: Rowan Tree Press, 1984), 1–8.
64. Karen Hess, ed., *Martha Washington's Booke of Cookery* (New York: Columbia University Press, 1981), 53, 67; Edgar S. Maclay, ed., *Journal of William Maclay* (New York: Albert and Charles Boni, 1890), 137–138.
65. Richard Hooker, ed., *A Colonial Plantation Cookbook: The Receipt Book of Harriott Pinckney Horry, 1770* (Columbia: University of South Carolina Press, 1984), 6–7, 56, 89.
66. F.J.F. Schantz, "Domestic Life and Characteristics of the Pennsylvania-German Pioneer," as in Pennsylvania-German Society *Proceedings* 10 (1901):18.
67. Jefferson Williamson, *American Hotel* (New York: Arno Press, 1975 [c. 1930]), 213; George R. Stewart, *American Ways of Life* (New York: Doubleday, 1954), 106; J. P. Brissot de Warville, *New Travels in the United States* 1788 (Cambridge: Belknap Press of Harvard University Press, 1964), 303.
68. *Die geschickte Hausfrau*, as in William Woys Weaver, *Sauerkraut Yankees: Pennsylvania German Foods and Foodways* (Philadelphia: University of Pennsylvania Press, 1983), 91–100.
69. Jean-Anthelme Brillat-Savarin [M.F.K. Fisher, trans.], *The Physiology of Taste; or, Meditations on Transcendental Gastronomy* (Washington, D.C.: Counter-

point, 1986), 271, 305; Samuel Adams Drake, *Old Boston Taverns and Tavern Clubs* (Boston: Cupples, Upham, 1886), 65–66.

70. Unknown Boston newspaper, April 18, 1794, as quoted in Arthur W. Brayley, "An Originator of Soups," *Boston Cooking-School Magazine* 10 (April 1906):445–446; Maria Parloa, *Miss Parloa's Kitchen Companion* (Boston: Dana Estes, 1880), 121.
71. Alfred L. Carroll, "On Digestion and Food," *Harper's New Monthly Magazine* 39 (November 1869):894.
72. Amelia Simmons, *American Cookery*, facsimile (New York: Oxford University Press, 1958), 16; Amelia Simmons, *American Cookery with an Introduction by Karen Hess*, 2nd ed., facsimile (Bedford, Mass.: Applewood Books, 1996), 19–20, 22–23.
73. Mary Randolph, *The Virginia Housewife with Historical Notes and Commentaries by Karen Hess*, facsimile (Columbia: University of South Carolina Press, 1984), 17–30.
74. N.K.M. Lee, *The Cook's Own Book* (Boston: Munroe & Francis, 1832), 27–28, 54–55, 202–213.
75. Eliza Leslie, *Directions for Cookery* (Philadelphia: E. L. Carey & A. Hart, 1837), 15–41, 55–56.
76. Sarah Josepha Hale, *The Good Housekeeper* (Boston: Weeks, Jordan, 1839), 57–62.
77. Paul Bourget, *Outre-Mer: Impressions of America* (New York: Charles Scribner's Sons, 1895), 353; Frances M. Wiley, *Model Cook Book* (Troy, New York: E. H. Lisk, 1884), 13–24; Walter Marshall, *Through America* (Freeport, New York: Books for Libraries Press, [1972]), 40.
78. Emma P. Ewing, *Soups and Soup Making* (Chicago and New York: Fairbanks, Palmer, 1882), 7–10.
79. Marion Harlan [pseud. for Mary Virginia Terhune], "Soup Making," *Home-Maker* 2 (April 1889):53, 55.
80. Harlan, "Soup Making," 54.
81. Elizabeth Robins Pennell, *The Feasts of Autolycus: The Diary of a Greedy Woman* (London: John Lane; New York: Merriam, 1896), 80.
82. Pennell, *The Feasts of Autolycus*, 79–80, 177–178; Michael McKirdy, foreword, Elizabeth Robins Pennell, *My Cookery Books*, reprint (West Yorkshire, England: Moxon Press, 1983), n.p.
83. Olive Green [Myrtle Reed], *One Thousand Simple Soups* (New York: G. Putnam's Sons, 1907), 1–4.

2. THE ORIGINAL TOMATO

1. Andrew F. Smith, "The Making of the Legend of Robert Gibbon Johnson and the Tomatoe," *New Jersey History* 108 (Fall/Winter 1990):59–74; Andrew F. Smith, *The Tomato in America: Early History, Culture and Cookery* (Columbia: University of South Carolina Press, 1994), 3–8.
2. Bernardino de Sahagún, *General History of the Things of New Spain (Florentine Codex)*, as in Charles Dibble and Arthur J. O. Anderson, trans., *Book 8—Kings and Lords*, monograph no. 14, part 9 (Santa Fe, New

Mexico: School of American Research and University of Utah, 1961), 37–38.

3. Alonso de Molina, *Vocabulario en lengua castellana y mexicana* (Mexico: En Casa de Centonio de Spinosa, 1571), 159b.
4. Hospital de la Sangre, *Libros de recibos y gastos,* file no. 455, as in Michael Allen and Robert Benson, eds., *First Images of America: The Impact of the New World on the Old,* 2 vols. (Berkeley: University of California Press, 1976), 2:859.
5. Antonio Latini, *Lo scalco alla moderna* 2 vols. (Naples: Con licengadé superiorise privilegio, 1692–1694), 1:390, 444, 551; 2:55, 162; Rudolf Grewe, "The Arrival of the Tomato in Spain and Italy: Early Recipes," *Journal of Gastronomy* 3 (Summer 1987):67–83.
6. Vincenzo Corrado, *Il cuoco galante opera mecconica* (Naples: Nella Stamperia Raimondia, 1786), 158.
7. Lancelot Addison, *An Account of West Barbary* (Oxford, 1671), reprint in John Pinkerton, *A General Collection of . . . Voyages and Travels,* 17 vols. (London: n.p., 1814), 15:405; Thomas Shaw, "Travels or Observations Relating to Barbary" (London, 1757), reprint ibid., 15:601.
8. Charles Bryant, *Flora Diaetetica; or, History of Esculent Plants* (London: Printed for B. White, 1783), 213.
9. John Claudius Loudon, *An Encyclopedia of Gardening* (London: Longman, Rees, Orme, Brown, and Green, 1827), 679.
10. Benjamin Daydon Jackson, ed., *A Catalogue of Plants Cultivated in the Garden of John Gerard in the Years 1596–1599* (London: Privately printed, 1876), 14; John Gerard, *The Herball; or, Generall Historie of Plantes* (London: Printed for the author, 1597), 275–276.
11. Gerard, *The Herball,* 275–276.
12. William Salmon, *Botanologia: The English Herbal; or, History of Plants* (London: Printed by I. Dawks for H. Rhodes and J. Taylor, 1710), 29.
13. Letter from Peter Collinson to John Custis of Williamsburg, February 6, 1742/43, in E. G. Swem, "Brothers of the Spade: Correspondence of Peter Collinson, of London, and John Custis, of Williamsburg, Virginia, 1734–1746," *American Antiquarian Proceedings* 58 (April 1948):101–102.
14. Phillip Miller, *The Gardeners Dictionary: Containing the Methods of Cultivating and Improving all sorts of Trees, Plants, and Flowers, for the Kitchen, Fruit, and Pleasure Gardens, etc.,* 4th ed., 3 vols. (London: Printed for the author, 1754), n.p.; ibid., 7th ed. (London: Printed for the author, 1759), n.p.
15. [John Hill], *A Supplement to Mr. Chambers's Cyclopedia; or, Universal Dictionary of Arts and Sciences,* 2 vols. (London: Printed for W. Innys and J. Richardson, 1753), vol. 1, listed under "Lyc."
16. John G. Stedman, *Narrative of a Five Year Expedition against the Revolted Negroes of Surinam in Guiana on the Wild Coast of South America from the Years 1772–1777,* introduction and notes by R. A. Lier, 2 vols. (Holland: Imprint Society, 1971), 2:343.
17. Henry Phillips, *The Companion for the Orchard* (London: Henry Colburn and Richard Bentley, 1831), 225–227.

18. [Judith Montefiore], *The Jewish Manual; or, Practical Information in Jewish and Modern Cookery with a Collection of Valuable Recipes and Hints Relating to the Toilette* (London: T. & W. Boone, 1846), facsimile (New York: NightinGale Books, 1983), 10–11.
19. Charles Marshall, *An Introduction to . . . Gardening,* 2 vols. (Boston: Etheridge, 1799), 1:264–265; *Gleanings from the Most Celebrated Books on Husbandry, Gardening and Rural Affairs* (Philadelphia: James Humphreys, 1803), 194.
20. Hannah Glasse, *The Art of Cookery Made Plain and Easy* (London: Printed for the author, 1758), 341; Richard Briggs, *The New Art of Cookery* (Philadelphia: W. Spotswood, R. Campbell, and B. Johnson, 1792), 80.
21. *Encyclopedia Britannica,* 3rd ed. (Edinburgh: A. Bell & C. Macfarquhar, 1797), listed under "Solanum," 597–598.
22. A[lexander] Hunter, *Culina Famulatrix Medicinae; or, Receipts in Modern Cookery, with a Medical Commentary,* 4th ed. (York: T. Wilson and R. Spence, 1806), 179, 251.
23. A.F.M. Willich, *The Domestic Encyclopedia; or, A Dictionary of Facts and Useful Knowledge,* 3 vols. (London: Murry and Highley, 1802), 3:128.
24. William Cobbett, *The American Gardener; or, A Treatise* (Claremont, New Hampshire: Manufacturing Company, [1819]), 160–161.
25. Maria Eliza Rundell, *American Domestic Cookery* (Baltimore: Fielding Lucas, June 1819), 140.
26. Phillips, *The Companion for the Orchard,* 227.
27. Salmon, *Botanologia,* 29; George C. Rogers Jr., ed., *The Papers of Henry Laurens,* 14 vols. (Columbia: University of South Carolina Press, 1974), 4:359; George Andrews, *Well's Register: Together with an Almanack* (Charlestown: Robert Wells, 1774), 15, 17.
28. Edward Long, *The History of Jamaica; or, General Survey of the Ancient and Modern State of That Island with Reflections on its Situations, Settlements, Inhabitants, Climate, Products, Commerce, Laws and Government* (London: T. Lowndes, 1774), 773.
29. Charles Varlo, *A New System of Husbandry,* 2 vols. (Philadelphia: Printed for the author, 1785), 2:291; Charles Marshall, *An Introduction to . . . Gardening,* 2 vols. (Boston: Etheridge, 1799), 1:264–265; *Gleanings from the Most Celebrated Books on Husbandry, Gardening and Rural Affairs* (Philadelphia: James Humphreys, 1803), 194.
30. Harriott Pinckney Horry Papers, Receipt Book, 28, South Carolina Historical Society no. 39-19; Richard Hooker, ed., *A Colonial Plantation Cookbook: The Receipt Book of Harriott Pinckney Horry, 1770* (Columbia: University of South Carolina Press, 1984), 25, 89.
31. Thomas Jefferson, *Notes on the State of Virginia, Written in the Year 1781, Somewhat Corrected and Enlarged in the Winter of 1782* ([Paris]: n.p., [1785]), 70; Bernard M'Mahon, *A Catalogue of Garden, Grass, Herb, Flower, Roots* (Philadelphia: n.p., 1800).
32. Henry S. Randall, *The Life of Thomas Jefferson,* 3 vols. (New York: Derby & Jackson, 1858), 1:opposite 44; William Booth, *A Catalogue of Kitchen Garden Seeds and Plants* (Baltimore: G. Dobbin and Murphy, 1810), 4;

John Lithen, *Catalogue of Garden Seeds* (Philadelphia: 1800); M'Mahon, *A Catalogue of Garden, Grass, Herb, Flower, Roots*; Bernard M'Mahon, *The American Gardener's Calendar, Adapted to the Climates and Seasons of the United States* (Philadelphia: Printed by B. Graves for the author, 1806), 319; Grant Thorburn, *A Catalogue of Kitchen-Garden, Flower-Seeds, etc.* (New York: Southwick and Hardcastle, 1807).

33. A.F.M. Willich [James Mease, ed.], *The Domestic Encyclopedia,* 3 vols. (Philadelphia: William Young Birch and Abraham Small, 1804), 3:506.
34. Michael Floy, *A Catalogue of Ornamental Trees, Flowering Shrubs* (New York: Pelsue and Young, 1816), 22.
35. Approved Recipes, 1795, 77, cookery manuscript at the New York State Historical Association, Cooperstown, New York; Receipt Book of Sally Bella Dunlop and James Dunlop, cookery manuscripts at the New York State Historical Association, Cooperstown, New York; manuscript cookbook, Mrs. George Read, New Castle, 1813, Holcomb Collection, Historical Society of Delaware, Wilmington.
36. Richard Alsop, *The Universal Receipt Book; or, Complete Family Direction by a Society of Gentlemen in New York* (New York: I. Riley, 1814), 45.
37. Thomas Brown, "Account of the Lineage of the Brown Family," manuscript dated 1865, pp. 48–49, in the Ambler-Brown Family Papers, Duke University Special Collections Library, Durham, North Carolina.
38. Rubens Peale, *Memorandum and Events of His Life,* n.d., back of p. 4, in the Peale-Sellers Papers, American Philosophical Society, Philadelphia.
39. *Historical Magazine* 6 (March 1862):102; *Country Gentleman* 38 (December 25, 1873):820, (June 5, 1873):366.
40. Louis Eustache Ude, *The French Cook* (Philadelphia: Carey, Lea and Carey, 1828), 2–4, 7, 46–64, 96, 256–257.
41. Eliza Leslie, *Domestic French Cookery, Chiefly Translated from Sulpice Barué* (Philadelphia: Carey & Hart, 1832), 21–22, 32–33, 35, 37–38, 73–74.
42. Ude, *The French Cook*; [Louis Eustache Audot], *French Domestic Cookery* (New York: Harper & Brothers, 1846), 272–273.
43. *Country Gentleman* 19 (May 15, 1862):318.
44. *Country Gentleman* 19 (May 15, 1862):318; George Perot Macculloch's Ledger, manuscript at Macculloch Hall Historical Museum, Morristown, New Jersey.
45. Charles W. Casper, manuscript report on canning, Salem County Historical Society, dated March 13, 1906; *American Agriculturist* 46 (May 1887):210.
46. *Boston Tribune,* in *New England Farmer* 9 (August 6, 1830):20; *New England Farmer* 9 (August 27, 1830):45.
47. *Salem Union,* as quoted in *Morris County Whig* (Morristown, N.J.), August 12, 1835.
48. *New-Jersey Journal* (Elizabeth, N.J.), August 11, 1835; *Constitution, and Farmers' and Mechanics' Advertiser* (Woodbury, N.J.), August 11 and September 15, 1835.
49. *Morris County Whig* (Morristown, N.J.), August 12, 1835.
50. *Sentinel of Freedom* (Newark, N.J.), August 18, 1835; *Paterson Intelligencer* (Paterson, N.J.), August 19, 1835.

51. For instance, see *Eagle* (Newark, N.J.), September 18, 1835; *Freeman's Banner* (Salem, N.J.), April 13, 1836.
52. *Somerset Whig* (Somerville, N.J.), September 18, 1838, 3.
53. *New York Times and Commercial Intelligencer,* July 19, 1839; *Western Sun & General Advertiser,* April 27, 1839–January 4, 1840; *United States Gazette,* August 25, 1838; *Philanthropist,* October 2, 1838; *New York Sun,* September 20, 1839; *Connecticut Courant,* November 30, 1839; *New York Express,* July 15, 1839; Robert Buist, *The Family Kitchen Gardener* (New York: J. C. Riker, 1847), 126; *Cincinnati Daily Gazette,* August 17, 1840. For more about tomato pills, see Smith, *The Tomato in America,* 102–131.
54. Lettice Bryan, *The Kentucky Housewife* (Cincinnati: Stereotyped by Shepard & Stearns, 1841), 24–25, 426.
55. R[ussell] T[hatcher] Trall, *The New Hydropathic Cook-book* (New York: Fowlers and Wells, 1854), 191.
56. *American Farmer* 4 (May 1822):40; *Germantown Telegraph,* November 3, 1847, 4; *Country Gentleman* 19 (May 15, 1862):318; *Genesee Farmer* 1 (September 10, 1831):293; *Historical Magazine* 6 (January 1862):35–36; *New England Cultivator* 1 (September 1852):258; *Lancaster Farmer* 2 (March 1870):52; *Genesee Farmer* 5 (March 7, 1835):78; *Horticulturist* 4 (September 1849):422; *Boston Courier,* August 1, 1845, 2.
57. Dio Lewis, *Talks about People's Stomach* (Boston: Fields, Osgood, 1870), 147–151.
58. Will W. Tracy, *The Tomato Culture* (New York, Orange Judd, 1907), 7; *Eclectic Medical Journal of Pennsylvania* 9 (September-October 1871):286; *American Agriculturist* 37 (June 1878):208; *New York Tribune,* August 31, 1878, 2.
59. *New York Tribune,* October 26, 1878, 3.
60. Homer O. Hitchcock, "Cancer Not Caused by Tomatoes," *Annual Report of Michigan State Board of Health* 6 (1878):35–38; *Country Gentleman* 53 (August 9, 1888):600; *American Agriculturist* 48 (December 1889):648; *Good Housekeeping* 15 (December 1892):267–268.
61. *New England Kitchen Magazine* 1 (August 1894):230.
62. *Journal of the American Medical Association* 26 (January-June 1896):1255–1258.
63. *New York Tribune,* November 28, 1896, 6.
64. *Hall's Journal of Health* 28 (March 1881):101–102; *Gardener's Monthly* 28 (July 1886):207.
65. Mrs. E. E. Kellogg, *Science in the Kitchen* (Battle Creek, Mich.: Health Publishing, 1892), 282, 284, 416; Mrs. E. E. Kellogg, comp., *Healthful Cookery: A Collection of Choice Recipes for Preparing Foods, with Special Reference to Health* (Battle Creek, Mich.: Modern Medicine Publishing, 1904), 177–178, 182, 188–190; John Harvey Kellogg, *The New Dietetics: A Guide to Scientific Feeding in Health and Disease,* rev. ed. (Battle Creek, Mich.: Modern Medicine Publishing, 1927), 315–316, 681.
66. Almeda Lambert, *Guide for Nut Cookery* (Battle Creek, Mich.: Joseph Lambert, 1899), 97, 305.
67. *Indiana Farmer's Guide,* 2nd new ser. 34 (October 14, 1922):1078; *Market Growers Journal* (August 15, 1923):118.

68. Anna Lindlahr and Henry Lindlahr, *The Lindlahr Vegetarian Cook Book and A B C of Natural Dietetics,* 15th ed. (Chicago: Lindlahr Publishing, 1922), 25.
69. W. C. Latta, *Outline History of Indiana Agriculture* (Lafayette, Ind.: Alpha Lambda Chapter of Epsilon Sigma Phi, 1938), 264; Della Lutes, *The Country Kitchen* (Boston: Little, Brown, 1936), 21.
70. While there are many reasons for identifying the tomato as a healthy vegetable, the specific attributes that have generated the latest interest are the carotenoids, a family of pigments found in yellow, orange, and red vegetables and green leafy vegetables. The human body cannot manufacture carotenoids; some studies have offered evidence that foods rich in them may play a role in preventing heart disease and cancer. There are over six hundred carotenoids, but the predominant one in tomatoes—and found almost exclusively in them—is lycopene, which gives tomatoes their red color; when lycopene is not present, the tomato is green or yellow. Lycopene was discovered in 1876 by the French botanist Alexis Millardet, who named it solanorubin. It was rediscovered in 1903 by British botanist C. A. Schunck, who renamed it lycopene. Dr. Edward Giovannucci, an assistant professor in the Department of Nutrition at Harvard School of Public Health and the Department of Medicine at the Harvard Medical School, examined responses of fifty thousand participants in the Harvard University Professionals Health Study that began in 1986. He concluded that the consumption of four vegetables and fruits was associated with lower prostate cancer risk: three of the foods he cited were tomato sauce, tomatoes, and pizza; the risk of prostate cancer was one third lower in men who ate tomato-based products. In another study, in northern Italy, a high correlation was drawn between tomato consumption and the lack of cancers of the digestive tract; of the 2,700 respondents, those who consumed seven or more servings of raw tomatoes every week had 60 percent less chance of developing cancer of the colon, rectum, and stomach. While medical studies have not specifically linked consumption of tomato soup with these benefits, there is every reason to suspect that it has the same healthful qualities as do other processed tomato products, such as tomato sauce, which contain high levels of lycopene.
71. *Economical Cookery* (Newark, N.J.: Benjamin Olds, 1839), 62, 99, 102–103, 108–109.
72. P. Thornton, *The Southern Gardener and Receipt Book* (Newark, N.J.: A. L. Denis, 1845), 55, 136, 141–142, 155–156, 163, 169.
73. *Cultivator,* 2nd ser. 4 (August 1847):247.
74. Edmund Morris, *Ten Acres Enough* (New York: J. Miller, 1864), 118, 156–157.
75. James C. Bonner, *A History of Georgia Agriculture, 1732–1860* (Athens: University of Georgia Press, 1964), 169; Fearing Burr, *The Field and Garden Vegetables of America* (Boston: Crosby and Nichols, 1863), 643; Morris, *Ten Acres Enough,* 118–119.
76. Patrick Neill, *The Fruit, Flower and Kitchen Garden* (Philadelphia: Henry

Carey Baird, 1851), 236–237; *Philadelphia Florist* 1 (May 1852):30; Morris, *Ten Acres Enough,* 128; *Southern Planter* 18 (February 1858):89.

77. *Annual Report of the Secretary of Agriculture* (Washington, D.C.: USDA, 1865), 269–270.
78. *American Agriculturist* 27 (August 1868):282.
79. *Horticulturist* 28 (October 1873):315.
80. *American Agriculturist* 7 (May 1848):137–138; *Genesee Farmer* 1 (July 1831):233.
81. *Horticulturist* 10 (March 1, 1855):130–134.
82. *U.S. Gazette,* as in *American Farmer,* 3rd ser. 3 (October 1841):181; Charles Wilkes, *Narrative of the U.S. Exploring Expedition,* 5 vols. (London: Wiley and Putnam, 1845), 3:309, 335; Nelson Klose, *America's Crop Heritage: The History of Foreign Plant Introduction by the Federal Government* (Ames: Iowa State College Press, 1950), 29.
83. Burr, *The Field and Garden Vegetables of America,* 643–654; ibid. (Boston: J. E. Tilton, 1865), 628–642.
84. *American Gardening* 16 (June 5, 1895), 214–215.
85. Henry A. Dreer, *Dreer's Garden Calendar* (Philadelphia: 1864), 21; Albert Emerson Benson, *History of the Massachusetts Horticultural Society* (Norwood, Mass.: Plimpton Press, 1929), 140; Burr, *The Field and Garden Vegetables of America* (Boston: J. E. Tilton, 1865), 639; *Landreths' Rural Register and Almanac* (Philadelphia, 1868), 56, 62–63; *Tilton's Journal of Horticulture* 4 (November 1868):276; Peter Henderson, *Peter Henderson's & Co. Abridged List of Everything for the Garden* (New York: 1876), 15; Liberty Hyde Bailey, "Notes on Tomatoes," Agricultural College of Michigan Bulletin no. 31 (Lansing: Thorp & Godfrey State Printers and Binders, 1887), 20; *Landreths' Seed Catalogue* (Philadelphia: 1887), 57; *American Gardening* 16 (June 5, 1895):214–215.
86. *Genesee Farmer* 17 (May 1856):156; Liberty Hyde Bailey, *The Survival of the Unlike* (New York: Macmillan, 1897), 481; U. P. Hedrick, ed., *Sturtevant's Edible Plants of the World* (New York: Dover, 1972), 460.
87. Robert Buist, *Buist's Garden Guide and Almanac* (Philadelphia: 1890), 101–103; Liberty Hyde Bailey, "The Origin of the Tomato; from a Morphological Standpoint," *American Garden and Floral Cabinet* 8 (April 1887):23.
88. *Livingston's Seed Annual* (Columbus, Ohio: A. W. Livingston's Sons, 1887), 44–45, and (1890), 11.
89. Alexander Livingston, *Livingston and the Tomato with a Preface and Appendix by Andrew F. Smith,* reprint (Columbus: Ohio University Press, 1998), xxxi, 142–143.
90. *American Agriculturist* 48 (February 1889):81; *Garden and Forest* 5 (October 12, 1892):487–488.
91. *American Agriculturist* 48 (February 1889):81, (March 1889):123; *Garden and Forest* 2 (January 2, 1889):5, 4 (February 11, 1891):68, 2 (September 4, 1889):432.
92. *Illustrated Annual Register of Rural Affairs* 6 (1870):187.
93. *American Grocer* 2 (May 23, 1870):231.
94. *American Grocer* 2 (May 23, 1870):231–232.

3. THE WELL-PRESERVED TOMATO

1. Robert P. Multhauf, *Neptune's Gift: A History of Common Salt* (Baltimore: Johns Hopkins University Press, 1978), 1–12.
2. Charles G. Beddows, "Fermented Fish and Fish Products," in Brian B. J. Wood, ed., *Microbiology of Fermented Foods,* 2 vols. (London and New York: Elsevier Applied Science Publishers, 1985), 2:2.
3. John Sinclair, *Code of Health and Longevity,* 4 vols. (Edinburgh: Printed for Arch. Constable, 1807), 1:432–436.
4. Sinclair, *Code of Health and Longevity,* 1:432–436.
5. [Nicholas] Appert, [K. G. Bitting, trans.], *The Book for All Households; or, The Art of Preserving Animal and Vegetable Substances for Many Years* (Chicago: n.p., 1920), 6; W. V. Cruess, *Commercial Fruit and Vegetable Products: A Textbook for Student, Investigator and Manufacturer,* 2nd ed. (New York and London: McGraw-Hill, 1938), 31–32; H. G. Muller, "Industrial Food Preservation in the Nineteenth and Twentieth Century," in C. Anne Wilson, ed., "Waste Not, Want Not," *Fourth Leeds Symposium on Food History and Traditions* (Edinburgh: Edinburgh University Press, 1991), 123–124.
6. Katherine Golden Bitting, "Un Bienfaiteur de l'humanité," in *A Complete Course in Canning,* 5th ed. (Baltimore: Canning Trade, 1924), 9–12; Edward F. Keuchel Jr., *The Development of the Canning Industry in New York State to 1960,* Ph.D. dissertation, Cornell University, 1970, 1–3.
7. Appert, *The Book for All Households,* xiii.
8. Appert, *The Book for All Households,* 9–11, 30–107; Grimod de la Reynière, *Almanach des gourmands* (Paris, 1805), as in Bitting, "Un Bienfaiteur de l'humanité," 11; Keuchel, *The Development of the Canning Industry in New York State to 1960,* 1–3.
9. Thomas Saddington, "Preserving Fruit without Sugar," *Journal of the Franklin Institute* n.s. 12 (August 1833):136–137; Mary B. Sim, *Commercial Canning in New Jersey: History and Early Development* (Trenton: New Jersey Agricultural Society, 1951), 3.
10. Sim, *Commercial Canning in New Jersey,* 3–4.
11. Appert, *The Book for All Households,* xxi–xxiv, 5–6; Keuchel, *The Development of the Canning Industry in New York State to 1960,* 5.
12. Muller, "Industrial Food Preservation in the Nineteenth and Twentieth Century," 126.
13. M. Appert, *The Art of Preserving* (London: Black, Perry, and Kingsbury, 1811; New York: D. Longworth, 1812); Bitting, "Un Bienfaiteur de l'humanité," 13.
14. Appert, *The Book for All Households,* 56–57.
15. Joseph Jerome Lalande [De la Lande], *Voyage en Italie,* 2nd ed., 9 vols. (Paris: Chez la Veuve Desaint, 1786), 1:510; *Le patrimoine culinaire de la France: Provence-Alpes-Côte d'Azur* (Paris: Albin Michel/CNAC, 1995), 302–304; *Le patrimoine culinaire de la France: Aquitaine* (Paris: Albin Michel/CNAC, 1997), 197–198; *Le patrimoine culinaire de la France: Languedoc-Roussillon* (Paris: Albin Michel/CNAC, 1998), 237–238.

16. J. C. Drummond and W. R. Lewis, "The Examination of Some Tinned Foods of Historic Interest," *Chemistry and Industry* (August 27, 1938):809–810; W. V. Cruess, *Commercial Fruit and Vegetable Products: A Textbook for Student, Investigator and Manufacturer,* 2nd ed. (New York and London: McGraw-Hill, 1938), 32–33; Muller, "Industrial Food Preservation in the Nineteenth and Twentieth Century," 126.
17. Muller, "Industrial Food Preservation in the Nineteenth and Twentieth Century," 127–128.
18. Muller, "Industrial Food Preservation in the Nineteenth and Twentieth Century," 127.
19. Lyman Underwood, "Incidents in the Canning Industry of New England," in Arthur I. Judge, ed., *A History of the Canning Industry: Souvenir of the Seventh Annual Convention of the National Canners' Association and Allied Associations* (Baltimore: Canning Trade, 1914), 12–13; Sim, *Commercial Canning in New Jersey,* 14.
20. Underwood, "Incidents in the Canning Industry of New England," 13.
21. *Horticultural Register* 4 (April 1, 1838):127.
22. Cruess, *Commercial Fruit and Vegetable Products,* 34–35.
23. Testimony in *John W. Jones* v. *Louis McMurray,* as cited in *American Grocer* 31 (February 4, 1884):9.
24. *Merchants' Review* 24 (August 14, 1891):7–8; *American Grocer* 30 (December 13, 1883):1960, 31 (February 4, 1884):9, 48 (July 20, 1892):8; *New York Tribune,* as reported in David Bishop Skillman, *The Biography of a College,* 2 vols. (Easton, Pa.: Lafayette College, 1932), 1:179–180.
25. *American Grocer* 95 (January 5, 1916):6–7.
26. *Bangor Whig and Courier,* as reprinted in *Maine Farmer* 17 (October 11, 1849):n.p.
27. *Southern Cultivator* 11 (August 1853):242.
28. Richard J. Hooker, *A History of Food and Drink in America* (Indianapolis and New York: Bobbs-Merrill, 1981), 207.
29. Herbert G. Schmidt, *Rural Hunterdon: An Agricultural History* (New Brunswick, N.J.: Rutgers University Press, 1945), 280; Hooker, *A History Food and Drink in America,* 214.
30. *Country Gentleman* 32 (December 3, 1868):376; *Horticulturist* 23 (November 1868):321–325.
31. U.S. Patent no. 12,153 to Robert Arthur, issued January 2, 1855; *Godey's Lady's Book* 55 (June 1857):566; *Fresh Fruits and Vegetables All Year at Summer Prices; and How to Obtain Them* (Philadelphia: Arthur, Burnham & Gilroy, 1857), 32–34.
32. Julian H. Toulouse, *Fruit Jars* (Camden, New Jersey: Thomas Nielson & Sons, and Everybodys Press, 1969), 340–347; E. F. Haskell, *The Housekeeper's Encyclopedia* (New York: D. Appleton, 1861), 131.
33. *Country Gentleman* 28 (August 1, 1866):115.
34. *Country Gentleman* 28 (August 1, 1866):115.
35. *American Grocer* 31 (February 14, 1884):9.
36. *Merchants' Review* 24 (August 14, 1891):7–8.
37. Hugh S. Orem, "Baltimore: Master of the Art of Canning," in Arthur I.

Judge, ed., *A History of the Canning Industry: Souvenir of the Seventh Annual Convention of the National Canners' Association and Allied Associations* (Baltimore: Canning Trade, 1914), 10.

38. *Business Review of the Counties of Hunterdon, Morris and Somerset, New Jersey* (Philadelphia: Pennsylvania Publishing, 1891), 15; Schmidt, *Rural Hunterdon*, 127.
39. Edmund Morris, *Ten Acres Enough* (New York: J. Miller, 1864), 120–161.
40. *Country Gentlemen* 56 (September 24, 1891):776; Charles W. Casper, untitled report on canning in Salem County, unpublished manuscript at the Salem County Historical Society, dated March 13, 1906, n.p.
41. *American Grocer* 14 (July 10, 1875):31.
42. *Report of the New Jersey Centennial Commissioners*, as cited in Sim, *Commercial Canning in New Jersey*, 85, 149; Scrap Book of Philadelphia Centennial Exposition, 1876, at the Prints and Pictures Collection at the Free Public Library of Philadelphia. The South Jersey Packing Company was acquired by John E. Diament in 1880; he eventually changed the name to the John E. Diament Company, which operated for over seventy years.
43. George Moore, "Farm Record 1853–77," unpublished handwritten manuscript at the Alexander Library (Special Collections), Rutgers University, New Brunswick, New Jersey; *American Grocer* 15 (April 15, 1876):424 and 16 (July 22, 1876):102, (July 29, 1876):130, (August 19, 1876):227, (August 26, 1876):274, (September 2, 1876):31, (September 9, 1876):344, (September 16, 1876):356.
44. *Maryland Farmer* 16 (August 1879):261.
45. *Scientific American* 2nd ser. 44 (April 30, 1881):281.
46. *Sunbeam* (Salem, New Jersey), January 27, 1888, 3.
47. *Business Review of the Counties of Hunterdon, Morris and Somerset, New Jersey* (Philadelphia: Pennsylvania Publishing, 1891), 15.
48. *Country Gentleman* 56 (September 24, 1891):776.
49. *Country Gentleman* 56 (September 24, 1891):776.
50. *Gardener's Chronicle* nv (May 31, 1879):687; *New York Times*, October 11, 1887, 14.
51. *New York Times*, October 11, 1887, 14; *Country Gentleman* 56 (September 24, 1891):776.
52. *Country Gentleman* 56 (September 24, 1891):776.
53. *American Gardening* 11 (April 16, 1890):238–239.
54. Charles W. Casper, untitled report on canning in Salem County, unpublished manuscript at the Salem County Historical Society, dated March 13, 1906, n.p.
55. *New York Times*, October 11, 1887, 14.
56. Charles W. Casper, untitled report on canning in Salem County, unpublished manuscript at the Salem County Historical Society, dated March 13, 1906, n.p.
57. *New York Times*, April 10, 1884, 6.
58. *American Grocer* 33 (January 1, 1885):14.
59. *American Grocer* 85 (January 4, 1911):7.

60. *New England Farmer* 12 (March 1860):146; *Gardening* 11 (September 1, 1903):376.
61. H. B. Meyer, ed and comp., *Journal of Proceedings of the Seventh Annual Convention of the National Association of State Dairy and Food Departments* (St. Paul, Minn.: n.p., 1903), 6.
62. *Baltimore Sun,* November 27, 1904, 16; *Philadelphia Record,* February 15, 1905.
63. House Committee on Interstate and Foreign Commerce, *Pure Food Hearings,* February 13–27, 1906, 3–67.
64. Mark Sullivan, *Our Times in the United States, 1900–1925,* 3 vols. (New York: Charles Scribner's Sons, 1929), 2:507.
65. *Country Gentleman* 32 (December 3, 1868):376; *Horticulturist* 23 (November 1868):321–325; *Maryland Farmer* 16 (August 1879):261; "Report of the Hunterdon County Agricultural Society," as in *Country Gentleman* 49 (August 21, 1884):696; *Historical Statistics of the United States: Colonial Times to 1970,* 2 vols. (Washington, D.C.: U.S. Department of Commerce, Bureau of the Census, 1975), 2:690–691; *New York Tribune,* as in *Scientific American Supplement* 42 (November 26, 1896):17,435.
66. *Grocer's Criterion* 31 (October 31, 1904):3; "Annual Tomato Packs: Table No. 1, 1891–1905," and "Annual Tomato Packs: Table No. 2, 1906–1913," in Arthur I. Judge, ed., *A History of the Canning Industry: Souvenir of the Seventh Annual Convention of the National Canners' Association and Allied Associations* (Baltimore: Canning Trade, 1914), 113; Earl Chapin May, *The Canning Clan: A Pageant of Pioneering Americans* (New York: Macmillan, 1937), 291–292.
67. May, *The Canning Clan,* 291–292.
68. May, *The Canning Clan,* 292; "Annual Tomato Packs: Table No. 1, 1891–1905," and "Annual Tomato Packs: Table No. 2, 1906–1913," in Arthur I. Judge, ed., *A History of the Canning Industry: Souvenir of the Seventh Annual Convention of the National Canners' Association and Allied Associations* (Baltimore: Canning Trade, 1914), 113, 121.
69. *Tinned Foods and How to Use Them* (London: Ward, Lock & Bowden, 1893), 19; *The Everyday Soup Book: A Recipe for Every Day of the Year including February 29th by G.P.* (London: Stanley Paul, 1915), 26–27.

4. THE TOMATO IN THE SOUP

1. John Conrade Cooke, *Cookery and Confectionary* (London: B. Bensley for W. Simpkin and R. Marshal, 1824), 11.
2. Ester Copley, *The Housekeeper's Guide* (London: Jackson and Walford, 1834), 177–178.
3. [James Dekay], "Turkish Preparation of the Tomato," *New York Farmer* 7 (November 1834):323–324.
4. *Pennsylvania Inquirer and Daily Courier,* August 20, 1835, 1.
5. [N.K.M. Lee], *The Cook's Own Book and Housekeeper's Register* (Boston: Munroe & Francis, 1832), 222.

6. Edward James Hooper, *The Practical Farmer, Gardener and Housewife* (Cincinnati: Geo. Conclin, 1839), 495.
7. Hooper, *The Practical Farmer, Gardener and Housewife*, 495–496.
8. Lettice Bryan, *The Kentucky Housewife* (Cincinnati: Stereotyped by Shepard & Stearns, 1841), 24–25, 426.
9. Maria Eliza Rundell, *A New System of Domestic Cookery Formed upon Principles of Economy*, ed. Emma Roberts, 66th ed. (London: John Murray, 1842), 18.
10. [Judith Montefiore], *The Jewish Manual; or, Practical Information in Jewish and Modern Cookery with a Collection of Valuable Recipes and Hints Relating to the Toilette* (London: T. & W. Boone, 1846), facsimile with introduction by Chaim Raphael (New York: NightinGale Books, 1983), 10–11, 97; [Louis Eustache Audot], *French Domestic Cookery* (New York: Harper & Brothers, 1846), 57.
11. Gentleman's Ordinary, menu for Birch's U.S. Hotel, Washington, D.C., dated July 11, 1846, in the menu collection at the New-York Historical Society.
12. Menu for Dinner to Celebrate Opening of the Revere House, Revere House, Boston, dated May 19, 1847; Bill of Fare, Cambridge Market Hotel, Cambridge, Massachusetts, dated July 21, 1857; Bill of Fare, Parker House, Boston, dated June 26, 1856, and Bill of Fare, Parker House, Boston, dated January 4, 1858, in the menu collection at the New-York Historical Society.
13. Menu for Fifth Anniversary Dinner of the New England Society of Louisiana, dated December 22, 1846, as in Martha Ann Peters, "The Saint Charles Hotel: New Orleans Social Center, 1837–1860," *Louisiana History* 1 (Summer 1960):202.
14. Bill of Fare, Irving House, New York, dated October 12, 1852, in the menu collection at the New York Public Library; Bill of Fare, Dey Street House, New York City, dated 1860, in the menu collection at the New-York Historical Society.
15. Bill of Fare, Matteson Hotel, Chicago, dated October 17, 1852, in the menu collection at the New-York Historical Society.
16. Menu for the Centennial Festival Banquet of the Saint Andrew's Society, at the Masonic Temple, Boston, dated November 29, 1856, in the menu collection at the New-York Historical Society.
17. Menu for the Centennial and Anniversary of the Burns Club in Honor of the Birth-Day of Robert Burns, Parker House, Boston, dated June 26, 1856, and Bill of Fare, Parker House, Boston, dated January 25, 1859, in the menu collection at the New-York Historical Society.
18. Bill of Fare for the Ladies' Ordinary, U.S. Hotel, Boston, dated January 6, 1851, and bills of fare for dinners at the Revere House in Boston: Anniversary of the Birth of Daniel Webster, dated January 18, 1856; New England Society for the Promotion of Manufactures and Mechanic Arts, dated January 30, 1856, and January 13, 1858; American Insurance Company, dated February 14, 1856; in Honor of the Election of Hon. A. P. Banks, for Speaker of the National House of Representatives, dated February 28, 1856; City Government of Salem, dated September 17,

1856; Connecticut Association, dated January 14, 1857; Revere House Company, dated January 22, 1857; Franklin Insurance Company, dated March 9, 1857, and March 8, 1858; City Council of Boston in Honor of Rear Admiral Mehemet Pasha and Suite, dated May 25, 1858; Neptune Insurance Company, dated October 30, 1858; Class of 1829, dated October 30, 1859; Festive Gathering of "The Saints," dated February 1, 1859; Banquet Given in Honor of National Typographical Union by the Boston Printers Union, dated May 5, 1859; Dinner Given by the Boston Chess Club in Honor of Paul Murphy, dated May 31, 1859. All in the menu collection at the New-York Historical Society.

19. Menu for Annual Dinner to the Independent Corps of Cadets at Parker's Restaurant, Boston, dated January 6, 1858, in the menu collection at the New-York Historical Society.
20. Bill of Fare, Stetson House, Long Branch, New Jersey, dated September 6, 1866, in the menu collection of the New York Public Library.
21. Felix Déliée, *The Franco-American Cookery Book* (New York: G. P. Putnam's Sons, 1884), 391, 434, 458, 474, 503, 540.
22. Thomas J. Murrey, *Fifty Soups* (New York: White, Stokes & Allen, 1884), 7, 15–16, 19–20, 36.
23. Charles Ranhofer, *The Epicurean* (New York: R. Ranhofer, 1894), 248, 264, 266, 267, 270.
24. Oscar Tschirky, *The Cook Book by "Oscar" of the Waldorf* (Chicago and New York: Werner, 1896), 22, 27–28, 34, 39, 41–42, 45.
25. *Good Housekeeping* 35 (August 1902):204.
26. [Lafcadio Hearn], *La Cuisine Creole: A Collection of Culinary Recipes from Leading Chefs and Noted Creole Housewives, Who Have Made New Orleans Famous for Its Cuisine* (New York: Will H. Coleman, 1885), 252; *Good Housekeeping* 3 (May 29, 1886):43.
27. Tschirky, *The Cook Book by "Oscar" of the Waldorf*, 11–12.
28. Janet M. Hill, "Seasonable Recipes," *Boston Cooking-School Magazine* 15 (April 1910):82.
29. David Paul Larousse, *The Soup Bible* (New York: John Wiley & Sons, 1997), 205–206.
30. Frances Willey, *The Model Cook Book* (Troy, New York: E. H. Lisk, 1884), 19; *Gardener's Monthly* 28 (July 1886):207; Almeda Lambert, *Guide for Nut Cookery* (Battle Creek, Mich.: Joseph Lambert, 1899), 306.
31. Celestine Eustis, *Cooking in Old Creole Days* (New York: R. H. Russell, 1904), 4–5.
32. Janet M. Hill, "Seasonable Recipes," *Boston Cooking-School Magazine* 14 (March 1910):378; Janet M. Hill, "Seasonable and Tested Recipes," *American Cookery* 20 (February 1916):529, (April 1916):690, 21 (June-July 1917):34.
33. Ida C. Bailey Allen, *Mrs. Allen on Cooking, Menus, Service* (Garden City, N.Y.: Doubleday, Page, 1924), 129, 132.
34. Rena Franklin, *Soups of Hakafri Restaurant* (Gainesville, Fla.: Triad, 1981), 109.
35. Mary Ella Milham, ed., *Platina: On Right Pleasure and Good Health* (Tempe, Ariz.: Medieval & Renaissance Texts & Studies, 1998), 295; [Lee], *The*

Cook's Own Book and Housekeeper's Register, 54; Fannie Merritt Farmer, *What to Have for Dinner* (New York: Dodge Publishing, 1905), 122; Janet M. Hill, "Seasonable Recipes," *Boston Cooking-School Magazine* 14 (October 1909):137–138.

36. *Presbyterian Cook Book* (Dayton, Ohio: Crooke & Co., 1873), 19–20; Maria Parloa, *Miss Parloa's Kitchen Companion* (Boston: Dana Estes, 1880), 141; Farmer, *What to Have for Dinner,* 39.
37. [Elizabeth Smith Miller], *In the Kitchen* (Boston: Lee and Shepard; New York: Lee, Shepard, and Dillingham, 1875), 49–50; Déliée, *The Franco-American Cookery Book,* 343, 391, 540; Juliet Corson, *Miss Corson's Cooking Manual of Practical Directions for Economical Every-Day Cooking* (New York: Dodd, Mead, 1877), 26; Almeda Lambert, *Guide for Nut Cookery* (Battle Creek, Mich.: Joseph Lambert, 1899), 304–305; Janet Hill, "Seasonable and Tested Recipes," *American Cookery* 21 (January 1917):449–450; M.R.L. Sharpe, *The Golden Rule Book: Six Hundred Recipes for Meatless Dishes* (Cambridge: University Press, 1908), 48, 67.
38. Mary Randolph, *The Virginia Housewife with Historical Notes and Commentaries by Karen Hess,* facsimile (Columbia: University of South Carolina Press, 1984), 31, 33–35, 107.
39. Mary Randolph, *The Virginia Housewife with Historical Notes and Commentaries by Karen Hess,* facsimile (Columbia: University of South Carolina Press, 1983), xxxii, 107; John Ayto, *Food and Drink from A to Z: A Gourmet's Guide,* 2nd ed. (Oxford and New York: Oxford University Press, 1993), 142; William Woys Weaver, "Additions and Corrections to Lowenstein's Bibliography of American Cookery Books, 1742–1860," *Proceedings of the American Antiquarian Society* 92 (1982):363–377; *Novisimo arte de cocina* (Philadelphia: Estereotipado é impreso por Compania, 1845), 72. I am indebted to Dan Strehl, senior librarian, Los Angeles Public Library, for locating a reference to an even earlier Spanish-language cookbook published in the United States: *Arte nuevo de cocina y reposteria. Acomodo al uso mexicano* (New York: en Casa de Lamuza, Mendía y C., Impresores libreros, 1828). Unfortunately, a copy of this work has not been located.
40. Karen Hess, "Historical Glossary," in Randolph, *The Virginia Housewife,* 275–276.
41. Thomas Jefferson, *Notes on the State of Virginia, Written in the Year 1781, Somewhat Corrected and Enlarged in the Winter of 1782* ([Paris]: n.p., [1785]), 70.
42. Randolph, *The Virginia Housewife,* 34–35, 275–277.
43. Eliza Leslie, *Directions for Cookery* (Philadelphia: E. L. Carey & A. Hart, 1837), 32–33.
44. Lettice Bryan, *The Kentucky Housewife* (Cincinnati: Stereotyped by Shepard & Stearns, 1841), 22–23.
45. P. Thornton, *The Southern Gardener and Receipt Book* (Newark, N.J.: A. L. Dennis, 1845), 141–142.
46. [Hearn], *La Cuisine Creole,* 18–22; *The Picayune Creole Cook Book,* 2nd ed. (New Orleans: Picayune, 1901), 34–38; *The Picayune Creole Cook Book,* 4th ed. (New Orleans: Picayune, 1910), 32–36.

47. *American Farmer* 12 (April 16, 1830):39; *New England Farmer* 8 (May 21, 1830):352; *New England Farmer* 9 (September 24, 1830):24.
48. [Sarah Rutledge], *The Carolina Housewife* (Charleston: W. R. Babcock, 1847), 43.
49. Will H. Coleman, comp., *Historical Sketch Book and Guide to New Orleans and Environs* (New York: Will H. Coleman, 1885), 86–91; Hodding Carter, *Lower Mississippi* (New York: Farrar & Rinehart, 1942), 418; *New Orleans as It Is* (Cleveland: W. W. Williams, [1885]), 56.
50. Blanche McManus, "Pedigree of the American Boiled Dinner," *American Cookery* 21 (February 1917):516; Ayto, *Food and Drink from A to Z*, 38.
51. Charles Elmé Francatelli, *A Plain Cookery Book for the Working Classes* (London: Routledge, Warne, and Routledge, 1852), 63–64; Pierre Blot, *Hand-Book of Practical Cookery* (New York: D. Appleton, 1867), 72–73.
52. André L. Simon, *A Concise Encyclopedia of Gastronomy,* reprint (Woodstock, N.Y.: Overlook Press, 1981), 315.
53. William Makepeace Thackeray, quoted in *The Picayune Creole Cook Book,* 2nd ed. (New Orleans: Picayune, 1901), 44.
54. William Makepeace Thackeray, "The Ballad of the Bouillabaisse," in *Ballads* (London: Smith, Elder, 1879), 59.
55. Déliée, *The Franco-American Cookery Book*, 451.
56. Murrey, *Fifty Soups*, 15.
57. Ranhofer, *The Epicurean*, 239–287; Grace Clergue and Gertrude Clergue, *Allied Cookery: British, French, Italian, Belgian, Russian* (New York and London: G. P. Putnam's Sons, 1916), 15; Florence Williams, "The Land of Bouillabaisse," *American Cookery* 31 (August-September 1926):96–97; Karl Schriftgiesser, *Oscar of the Waldorf* (New York: E. P. Dutton, 1943), 197.
58. *Boston Evening Post,* September 23, 1751, as cited in Richard J. Hooker, *The Book of Chowder* (Boston: Harvard Common Press, 1978), 27.
59. [Hannah] Glasse, *The Art of Cookery Made Plain and Easy* (Alexandria, Virginia: Cottom and Stewart, 1805), 262.
60. Randolph, *The Virginia Housewife*, 99; Thornton, *The Southern Gardener*, 138.
61. J. Chadwick, *Home Cookery: A Collection of Tried Receipts* (Boston: Crosby, Nichols, 1853), originally published with *The New Family Book; or, Ladies' Indispensable Companion,* reprint (Birmingham, Ala.: Oxmoor Press, 1984), 84–86.
62. Lydia Maria Child, *The Frugal Housewife* (Boston: March & Capen and Carter & Hendee, 1829), 31–32, 114–115; Hooker, *The Book of Chowder,* 9, 104.
63. P. P. Roger, *The Roger Cookery* (Boston: Joseph Dowe, 1838), 40–41.
64. Blot, *Hand-Book of Practical Cookery*, 159–160. I am indebted to Karen Hess, who brought this reference to my attention.
65. Parloa, *Miss Parloa's Kitchen Companion*, 161–162; Murrey, *Fifty Soups*, 19–20; Ranhofer, *The Epicurean*, 267; Fannie Merritt Farmer, *Boston Cooking-School Cook Book* (Boston: Little, Brown, 1896), 128.
66. Parloa, *Miss Parloa's Kitchen Companion*, 161–162; Murrey, *Fifty Soups*, 19–20; Ranhofer, *The Epicurean*, 267; Farmer, *Boston Cooking-School Cook Book*, 128; Sarah Tyson Rorer, *The Philadelphia Cook Book* (Philadelphia:

Arnold, 1886), 34; *Good Housekeeping Everyday Cook Book* (New York: Phelps Publishing, 1903), 286; Olive Green [Myrtle Reed], *One Thousand Simple Soups* (New York: G. Putnam's Sons, 1907), 260.

67. *Larkin Housewives' Cook Book: Good Things to Eat and How to Prepare Them* (Buffalo, Chicago, and Peoria: Larkin, 1915), 12; *Good Housekeeping's Book of Good Meals* (New York: Good Housekeeping, 1929), 131.
68. Jessup Whitehead, *Cooking for Profit: A New American Cook Book,* 3rd ed. (Chicago: Jessup Whitehead, 1893), 98–99; Joseph Vachon, *Vachon's Book of Economical Soups and Entreés* (Chicago: Hotel Monthly Press, 1903), 63–64.
69. Margaret Compton, *Grand Union Cook Book* (New York: Grand Union Tea Co., 1902), 51–52; *Pictorial Review Standard Cook Book* (New York: Pictorial Review, 1931), 71.
70. Virginia Elliot and Robert Jones, *Soups and Sauces* (New York: Harcourt, Brace, 1934), 21.
71. Ann Roe Robbins, *100 Summer and Winter Soups* (New York: Thomas Y. Crowell, 1943), 36.
72. Eleanor Early, *A New England Sampler* (Boston: Waverly House, 1940), 349–350; Hooker, *The Book of Chowder*, 9.
73. Louis P. De Gouy, *The Soup Book: 770 Recipes* (New York: Greenberg, 1949), 207, 220–221.
74. De Gouy, *The Soup Book*, 287–288, 291.
75. Archie Croydon Hoff, *Soups and Consommés* (Los Angeles: International Publishing, 1913), 47; Frances Troy Northcross, *250 Delicious Soups* (Chicago: Culinary Arts Institute, 1940), 22.
76. Larousse, *The Soup Bible*, 257; James Peterson, *Splendid Soups: Recipes and Master Techniques for Making the World's Best Soups* (New York: Bantam Books, 1993), 302–303.
77. [Estelle Woods Wilcox, comp.], *Centennial Buckeye Cook Book* (Marysville: J. H. Shearer & Son, 1876), 311; *Manual for Army Cooks Prepared under the Direction of the Commissary General of Subsistence* (Washington, D.C.: Government Printing Office, 1896), 187–188. I am indebted to Joseph Carlin for locating the "salza" recipe.
78. Encarnación Pinedo, *The Spanish Cook: A Selection of Recipes from Encarnación Pinedo's* El cocinero español, *Edited and Translated by Dan Strehl* (Pasadena: Weather Bird Press, 1992), 1–3. This is the first known Spanish-language cookbook published for use in the United States. As previously noted, the earlier Spanish-language cookbook, *Novisimo arte de cocina,* was printed in Philadelphia in 1845, but it appears to have been distributed solely in Mexico. See *Novisimo arte de cocina* (Philadelphia: Estereotipado é impreso por Compania, 1845), 72.
79. May E. Southworth, comp., *One Hundred and One Mexican Dishes* (San Francisco: Paul Elder, 1906), 4.
80. Victor Hirtzler, *The Hotel St. Francis Cook Book* (Chicago: Hotel Monthly Press, 1919), 221, 223, 296, 321, 334.
81. Gaetano Jerna, "Qualche cenno di storia sul pomodoro in Italia," *Humus* 3 (September 1947): 27; Joseph Jerome Lalande [De la Lande], *Voyage en Italie,* 2nd ed., 9 vols. (Paris: Chez la Veuve Desaint, 1786), 1:510–511.

82. Eliot Lord, John J. D. Trenor, and Samuel J. Barrows, *The Italian in America* (New York: B. F. Buck, 1905), 121–123; Richard J. Hooker, *A History of Food and Drink in America* (Indianapolis and New York: Bobbs-Merrill, 1981), 290–292; Nancy Verde Barr, *We Called It Macaroni: An American Heritage of Southern Italian Cooking* (New York: Alfred A. Knopf, 1996), 46.
83. *Presbyterian Cook Book* (Dayton, Ohio: Crooke, 1873), 20; Parloa, *Miss Parloa's Kitchen Companion*, 142; Tschirky, *The Cook Book by "Oscar" of the Waldorf*, 45.
84. The earliest American cookbooks with "Italian" noted in the titles are Charles Elmé Francatelli, *The Modern Cook: A Practical Guide to the Culinary Art in All its Branches, Comprising, in Addition to English Cookery, the Most Approved and Recherché System of French, Italian and German Cookery* (New Orleans: Thomas L. White, 1859); Lia Rand [Guilia Brandies], *The Philosophy of Cooking Comprising Forty-one Explanatory Letters and Three Hundred and Ten Foreign Recipes: French, German and Italian, Adapted for the American Home Table* (Brooklyn: H. G. Dougherty, 1894); Antonia Isola, *Simple Italian Cookery* (New York: Harper & Brothers, 1912); Julia Cuniberti, *Practical Italian Recipes for American Kitchens* (Janesville, Wisconsin: Gazette Printing, 1917); Jack Cusimano, *Economical Italian Cook Book* (Los Angeles: The author, 1917).
85. The earliest located recipe for minestrone in America dates to the 1870s and 1880s. See Lillie Hitchcock Coit [John C. Craig, ed.], *The Recipe Book of Lillie Hitchcock Coit, Introduction by Carol Hart Field* (Berkeley: University of California, 1998), 18.
86. J. E. Lighter, ed., *Random House Historical Dictionary of American Slang*, 3 vols. (New York: Random House, 1994), 1:249; Charles Elmé Francatelli, *Francatelli's Modern Cook* (Philadelphia: David McKay, [1895]), 135–136; Ayto, *Food and Drink from A to Z*, 37; John F. Mariani, *The Dictionary of American Food and Drink* (New York: Hearst Books, 1994), 35–36; Joyce Toomre, trans., *Classic Russian Cooking: Elena Molokhovets'* A Gift to Young Housewives (Bloomington and Indianapolis: Indiana University Press, 1992), 55; Darra Goldstein, *À la Russe: A Cookbook of Russian Hospitality* (New York: Random House, 1983), 155–159; Frances Troy Northcross, *250 Delicious Soups* (Chicago: Culinary Arts Institute, 1940), 7; Dallas Chapter of the Association of the Junior Leagues of America, ed. and comp., *The Junior League of Dallas Cook Book*, 2nd ed. (Dallas, Texas: n.p., ca. 1950), 38–39.
87. For discussions of charitable cookbooks, see Margaret Cook, *America's Charitable Cooks: A Bibliography of Fund-Raising Cook Books Published in the United States (1861-1915)* (Kent, Ohio: n.p., 1971), and Anne L. Bower, ed., *Recipes for Reading: Community Cookbooks, Stories, Histories* (Amherst: University of Massachusetts Press, 1997).
88. *Presbyterian Cook Book* (Dayton, Ohio: Crooke, 1873), 19–20.
89. Ladies of the Saint Francis Street Methodist Episcopal Church, South, Mobile, Alabama, comp., *Gulf City Cook Book* (Dayton, Ohio: United Brethren Publishing House, 1878), 5–7, 11–13, 17–18, 20.
90. *Par Excellence, Manual of Cookery* (Chicago: Saint Agnes Guild of the Church of the Epiphany, 1888), 9–10.

91. *Par Excellence, Manual of Cookery*, 9–10.
92. *The "Home" Cook Book*, 2nd ed. (Cinnaminson, N.J.: Children's Summer Home, 1915), 6; Minneapolis Public Library Staff Association, comp., *Library Ann's Cook Book* (Minneapolis, n.p., 1928), 43, 46–47.
93. *Ransom's Family Receipt Book* (Buffalo: D. Ransom, 1885), 3.
94. *Helps for the Hostess* (Camden, N.J.: Joseph Campbell Company, c. 1916), 18, 19, 34, 39, 41, 48–49.

5. SOUPY SALES

1. *The Everyday Soup Book: A Recipe for Every Day of the Year including February 29*th *by G. P.* (London: Stanley Paul, 1915), 26.
2. J.H.W. Huckins & Co.'s letter to Darwin W. Desmond, Newburgh, N.Y., dated March 20, 1888.
3. U.S. Patent no. 47,545 to James H. W. Huckins, issued May 2, 1865.
4. *Grocer* (Baltimore) 3 (July 1, 1876):17; Artemas Ward, *The Grocers' Hand-Book and Directory* (Philadelphia: Philadelphia Grocer Publishing, 1882), 208.
5. The quotations are noted in an advertising brochure for "Huckins' Soups" (Boston: J.H.W. Huckins & Co., n.d.), 1–3.
6. The quotations are noted in an advertising brochure for "Huckins' Soups."
7. *Good Housekeeping* 4 (February 19, 1887):vi, (March 5, 1887):viii, (March 19, 1887):i, (April 2, 1887):i, (April 30, 1887):311; "Huckins' Soups Wholesale Price List" (Boston: J.H.W. Huckins & Co., n.d.).
8. Letter by Darwin W. Esmond of Newburgh, N.Y., to Huckins & Company, Boston, dated March 20, 1888; *Par Excellence, Manual of Cookery* (Chicago: Saint Agnes Guild of the Church of the Epiphany, 1888), n.p.
9. Earl Chapin May, *The Canning Clan: A Pageant of Pioneering Americans* (New York: Macmillan, 1937), 339–340; Mary B. Sim, *Commercial Canning in New Jersey: History and Early Development* (Trenton: New Jersey Agricultural Society, 1951), 252–253.
10. May, *The Canning Clan*, 341.
11. Marion Harlan, "Soup Making," *The Home-Maker* 2 (April 1889):54.
12. Harlan, "Soup Making," 55.
13. Harlan, "Soup Making," 55–56.
14. A. Biardot, *Franco-American Soups: How They Are Made* (Jersey City Heights: Franco-American Food Company, n.d.), n.p.
15. U.S. Trademark no. 17,952 to Franco-American Food Company, issued May 27, 1890; *Modern Housekeeping and Food News* 9 (January 1907):39–40; *Kirmess Cook Book: A Collection of Well-tested Recipes from the Best Housekeepers of Jersey City and Elsewhere. Compiled for the Kirmess Given for the Benefit of Christ Hospital of Jersey City, Nov. 7–8–9–10, 1906* ([Jersey City?: n.p., ca. 1907]), 51–52.
16. *American Grocer* 23 (June 3, 1880):1427; *The Esculent for Advancement of the Best Eating* (Boston: Wood, Pollard & Co., Importers and Grocers, January 1906), 21–22.

17. Sim, *Commercial Canning in New Jersey*, 83; U.S. Trademark no. 29,447 to Joseph Campbell Preserve Company, issued January 12, 1897.
18. *West Jersey Press*, August 5, 1868, 2; George R. Prowell, *History of Camden County, New Jersey* (Philadelphia: L. J. Richards, 1886), 536–537; C. T. Lee, "A History of the Canned Meat Industry," in Arthur I. Judge, ed., *A History of the Canning Industry: Souvenir of the Seventh Annual Convention of the National Canners' Association and Allied Associations* (Baltimore: Canning Trade, 1914), 41; Sim, *Commercial Canning in New Jersey*, 83.
19. *American Grocer* 7 (November 30, 1872):942, (December 21, 1872):1087; Sim, *Commercial Canning in New Jersey*, 148.
20. *American Grocer* 9 (August 23, 1873):355; U.S. Trademark no. 1,661 to Anderson & Campbell, issued March 10, 1874; U.S. Trademark no. 29,447 to Joseph Campbell Preserve Company, issued January 12, 1897.
21. *American Grocer* 9 (August 23, 1873):355; *Philadelphia Directory*, 1874, as in Mary Hines, Gordon Marshall, and William Woys Weaver, *The Larder Invaded: Reflections on Three Centuries of Philadelphia Food and Drink* (Philadelphia: Library Company of Philadelphia and Historical Society of Pennsylvania, 1987), 81.
22. Sim, *Commercial Canning in New Jersey*, 149.
23. Sim, *Commercial Canning in New Jersey*, 85.
24. Sim, *Commercial Canning in New Jersey*, 88.
25. U.S. Trademark No. 4,885 to Abraham Anderson, issued July 17, 1877; Sim, *Commercial Canning in New Jersey*, 85–89.
26. *Tilton's Journal of Horticulture* 4 (November 1868):276; Robert Manning, *History of the Massachusetts Horticultural Society, 1829–1878* (Boston: Rand, Avery, 1880), 370; *Popular Gardening* 2 (January 1887):60.
27. *American Grocer* 46 (July 22, 1891):21.
28. Prowell, *History of Camden County, New Jersey*, 536–537; U.S. Trademark No. 4,885 to Abraham Anderson, issued July 17, 1877; *American Grocer* 45 (January 14, 1891):19; Sim, *Commercial Canning in New Jersey*, 85–89.
29. *Table Talk* 11 (June 1896):iv.
30. *American Grocer* 27 (March 9, 1882):562, 47 (June 15, 1892):25.
31. Prowell, *The History of Camden County, New Jersey*, 537; "Campbell's Soup," *Fortune* 12 (November 1935):71.
32. *Table Talk* 11 (January 1896):inside cover.
33. May, *The Canning Clan*, 341–342.
34. James H. Collins, *The Story of Canned Foods* (New York: E. P. Dutton, 1924), 186–187; Douglas Collins, *America's Favorite Food: The Story of Campbell Soup Company* (New York: Harry N. Abrams, 1994), 30.
35. May, *The Canning Clan*, 342.
36. Canning Trade, *A Complete Course in Canning*, 3rd ed. (Baltimore: Canning Trade, 1914), 155–156.
37. Clyde H. Campbell, *Campbell's Book: A Text Book on Canning, Preserving and Pickling* ([New York: Canning Age, c. 1929]), 49–50.
38. "The Story of Campbell's Soups . . . ," *Post-Telegram* (Camden, New Jersey), March 14, 1922, as in *Optimist* 11 (June 1, 1922):4; "Campbell Soup Company Historical Synopsis," undated mimeographed paper in the Campbell Soup Company archives, Camden, New Jersey.

39. Thomas Hine, *The Total Package: The Evolution and Secret Meanings of Boxes, Bottles, Cans, and Tubes* (New York: Little, Brown, 1995), 87.
40. Collins, *The Story of Canned Foods*, 187.
41. "Campbell's Soup," *Fortune* 12 (November 1935):72.
42. Sim, *Commercial Canning in New Jersey*, 152–156; Collins, *America's Favorite Food*, 41.
43. "Campbell Soup Company Historical Synopsis," undated mimeographed paper in the Campbell Soup Company archives, Camden, New Jersey, 3–4; Collins, *America's Favorite Food*, 41.
44. "Campbell Soup Company Historical Synopsis," undated mimeographed paper in the Campbell Soup Company archives, Camden, New Jersey, 4–5; John T. Dorrance, as quoted in Collins, *America's Favorite Food*, 32, 42.
45. Collins, *America's Favorite Food*, 37.
46. Sim, *Commercial Canning in New Jersey*, 157.
47. Sim, *Commercial Canning in New Jersey*, 157.
48. *Saturday Evening Post* (April 12, 1930):37; Irma S. Rombauer, *The Joy of Cooking*, facsimile (New York: Scribner, 1998), 19–20.
49. *Journal of the Proceedings of the Sixth Annual Convention of the National Association of State Dairy and Food Departments* (Portland, Oregon: 1902), 308.
50. *Good Housekeeping* 32 (April 1901):171; *American Grocer* 71 (June 8, 1904):12; *Modern Housekeeping and Food News* 9 (January 1907):39–40.
51. *Burt Olney's Soups, Salads and Desserts*, 8th ed. (Oneida, New York: Burt Olney Canning Company, 1910).
52. Advertisement for the T. A. Snider Preserve Company featuring "Snider's Tomato Specialties"; *Table Talk* 11 (December 1896):ii; *What to Eat* 15 (November 1903):168–69; *Ladies' Home Journal* 27 (June 1910):34.
53. *New Idea, Woman's Magazine* 9 (June 1904) 78; *Ladies' Home Journal* 33 (March 1916):103; May, *The Canning Clan*, 345; Eleanor Foa Dienstag, *In Good Company: 125 Years at the Heinz Table (1869–1994)* (New York: Warner Books, 1994), 246.
54. J. I. Holcomb, *Salesology of the Butter-Kist Popcorn Machine* (Indianapolis: Holcomb and Hoke Manufacturing Company, 1917), 136–137.
55. "Soup, Salesmen and Sales Promotion . . . They Run Together in the Campbell Sales Co.," *Printer's Ink* (February 18, 1955):32.
56. "Campbell Soup Company Historical Synopsis," undated mimeographed paper in the Campbell Soup Company archives, Camden, New Jersey, 9.
57. "Campbell Soup Company Historical Synopsis," undated mimeographed paper in the Campbell Soup Company archives, Camden, New Jersey, 8; "Campbell's Soup," *Fortune* 12 (November 1935):74–76; May, *The Canning Clan*, 344.
58. "Campbell Soup Company Historical Synopsis," undated mimeographed paper in the Campbell Soup Company archives, Camden, New Jersey, 10–11; "Campbell's Soup," *Fortune* 12 (November 1935):74–76; May, *The Canning Clan*, 344; Collins, *America's Favorite Food*, 119.
59. "Campbell Soup Company Historical Synopsis," undated mimeographed paper in the Campbell Soup Company archives, Camden, New Jersey,

10–11; Reay Tannahill, *Food in History,* rev. ed. (New York: Crown Publishers, 1989), 207.

60. Campbell Soup Company press release, dated November 18, 1998.
61. "Campbell's Soup," *Fortune* 12 (November 1935):69–70.
62. Frank Wright Pratt, *Boyhood Memories of Old Deerfield* (Portland, Maine: Southworth-Anthoensen Press, 1936), 244.
63. May, *The Canning Clan,* 304; *Indianapolis Star Magazine,* July 19, 1959, 26, 28.
64. Ingrid Nelson Waller, *Where There Is Vision: The New Jersey Agricultural Experiment Station, 1880–1955* (New Brunswick, N.J.: Rutgers University Press, 1955), 53.
65. *Indiana Farmer's Guide,* 2nd ser. 34 (October 14, 1922):1078; *Market Growers Journal* (August 15, 1923):118.
66. May, *The Canning Clan,* 304–306.
67. U.S. Patent no. 1,746,657 to Walter J. Kemp, issued February 11, 1930; May, *The Canning Clan,* 306.
68. "Campbell Soup Company Historical Synopsis," undated mimeographed paper in the Campbell Soup Company archives, Camden, New Jersey, 11.
69. "Campbell's Soup," *Fortune* 12 (November 1935):68.
70. Belle Terre Garden Club, *Belle Terre Favorites* (New York: Stewart, Warren & Benson, 1936), 13.
71. John Ayto, *Food and Drink from A to Z: A Gourmet's Guide,* 2nd ed. (Oxford and New York: Oxford University Press, 1993), 34; John Mariani, *The Dictionary of Italian Food and Drink* (New York: Broadway Books, 1998), 32–33.
72. *Journal of the Proceedings of the Sixth Annual Convention of the National Association of State Dairy and Food Departments* (Portland, Oregon, 1902), 90; Deposition of Charles Loudon, as in *American Food Journal* 4 (January 15, 1909):22; A. W. Carswell, memo, "V-8 Cocktail Vegetable Juices—History," dated December 18, 1953, in the Campbell Soup Company archives, Camden, New Jersey.
73. "Campbell's Soup," *Fortune* 12 (November 1935):68.
74. Martin Mayer, " 'We Think We Know More about the Tomato than Anyone Else in the World,' " *Cosmopolitan* 170 (May 1971):221.
75. Collins, *The Story of Canned Foods,* 188.
76. "Campbell Soup Company Historical Synopsis," undated mimeographed paper in the Campbell Soup Company archives, Camden, New Jersey, 4–9.
77. Holcomb, *Salesology of the Butter-Kist Popcorn Machine,* 136–137.
78. David Traxel, *1898: The Tumultuous Year of Victory, Invention, Internal Strife, and Industrial Expansion That Saw the Birth of the American Century* (New York: Alfred A. Knopf, 1998), 294.
79. "Campbell Soup Company Historical Synopsis," undated mimeographed paper in the Campbell Soup Company archives, Camden, New Jersey, 4–9.
80. Diane M. Goff and Geraldine Wine, "The Life and Art of Grace G. Drayton, Part 2," *Doll News* (Summer 1988):21.

81. *Saturday Evening Post* 191 (August 10, 1918):23; *Saturday Evening Post* 191 (July 27, 1918):23; *Good Housekeeping* 63 (August 1916):advertising section, 33; *Ladies' Home Journal* 27 (February 1910):40 and 28 (April 1911):1, (May 1911):33, (June 1911):1, (July 1911):38, (August 1911):49, (October 1911):77, (November 1911):43.
82. *Good Housekeeping* 55 (December 1912):advertising section, 22.
83. *Good Housekeeping* 63 (August 1916):advertising section, 33.
84. "Campbell's Soup," *Fortune* 12 (November 1935):126, 128; "Campbell Soup Company Historical Synopsis," undated mimeographed paper in the Campbell Soup Company archives, Camden, New Jersey, 11, 18; Effie Awards, http://www.effie.org/mhf94.htm.
85. Joseph Campbell Company, *Campbell's Menu Book* (Camden, N.J.: Joseph Campbell Company, 1910), n.p.
86. *Campbell's Condensed Tomato Soup* (Camden, N.J.: Joseph Campbell Company, 1914), n.p.
87. *Helps for the Hostess* (Camden, N.J.: The Joseph Campbell Company, ca. 1916), 18–19, 34–35, 38–39, 50.
88. *The Eagle Cook Book* (Brooklyn: Brooklyn Evening Eagle, 1915), n.p.
89. Rombauer, *The Joy of Cooking*, 19–20.
90. Rombauer, *The Joy of Cooking*, 26.
91. Belle Terre Garden Club, *Belle Terre Favorites*, 69; *Grays Harbor Orthopedic Auxiliary Cook Book* ([Aberdeen, Washington]: December 1936), 29; *Kenwood Methodist Cook Book* (Waukesha, Wisconsin: Freeman Printing, 1939), 42, 49, 81.
92. Carlton Lake, "How the Aspic from Topeka Won the Heart of Alice B. Toklas," *New York Times*, June 5, 1988; Alice B. Toklas, *The Alice B. Toklas Cookbook, Foreword by Maureen Duffy* reprint (London: Serif, 1994), 252–253.
93. *The Kitchen Directory, and American Housewife* (New York: Mark H. Newman, 1844), 44; Catherine Ester Beecher, *Miss Beecher's Domestic Receipt Book* (New York: Harper and Brothers, 1846), 78–79; L. G. Abell, *The Skilful* [sic] *Housewife's Book* (New York: C. M. Saxton, 1852), 104–105; [Julia C. Andrews], *Breakfast, Dinner and Tea* (New York: D. Appleton, 1859), 143; E. F. Haskell, *The Housekeeper's Encyclopedia* (New York: D. Appleton, 1861), 131; A. P. Hill, *Mrs. Hill's New Cook Book* (New York: Carleton Publishers, 1872), 365; E. E. Kellogg, *Science in the Kitchen* (Battle Creek, Mich.: Health Publishing, 1892), 293; Almeda Lambert, *Guide for Nut Cookery* (Battle Creek, Mich.: Joseph Lambert, 1899), 397–398. All cited newspapers are from the Campbell Soup Company archives clipping book, Camden, New Jersey.
94. *The New England Hotel Women's Cookbook* ([Boston]: New England Women's Hotel Relief Association, 1933), 42; Kay Morrow, comp. and ed., *The Western Cook Book: 250 Fine Recipes including Many Characteristic Northwestern, Southwestern and Mexican Dishes* (Reading, Pa.: Culinary Arts Press, 1936), 38. I am indebted to Joseph Carlin for the former citation.
95. Material prepared for the author by Jackie Finch, senior manager, Global Consumer Food Center, Campbell Soup Company, Camden, New Jersey.

96. Campbell Soup Company, *Wonderful Ways with Soups* (Camden, N.J.: Campbell Soup Company, 1958); Campbell Soup Company, *A Campbell's Cookbook: Cooking with Soup* ([Camden, N.J.]: Campbell Soup Company, 1962); *Campbell's Creative Cooking with Soup* (New York: Beekman House, 1985).
97. Effie Awards, http://www.effie.org/mhf94.htm.

6. SOUPER REVOLUTIONS

1. W. V. Cruess, *Commercial Fruit and Vegetable Products: A Textbook for Student, Investigator and Manufacturer,* 2nd ed. (New York and London: McGraw-Hill, 1938), 34–35.
2. W.H.H. Stevenson, "Cans and Can-Making Machinery," in Arthur I. Judge, ed., *A History of the Canning Industry: Souvenir of the Seventh Annual Convention of the National Canners' Association and Allied Associations* (Baltimore: Canning Trade, 1914), 92-93; Cruess, *Commercial Fruit and Vegetable Products*, 36.
3. Stevenson, "Cans and Can-Making Machinery," 93; Cruess, *Commercial Fruit and Vegetable Products*, 35–36.
4. Hugh S. Orem, "Baltimore: Master of the Art of Canning," in Arthur I. Judge, ed., *A History of the Canning Industry: Souvenir of the Seventh Annual Convention of the National Canners' Association and Allied Associations* (Baltimore: Canning Trade, 1914), 10-11; Cruess, *Commercial Fruit and Vegetable Products*, 36.
5. *Gardener's Chronicle,* n.v. (May 31, 1879):687; *New York Times*, October 11, 1887, 14; John D. Cox, "The Evolution of Tomato Canning Machinery," in Arthur I. Judge, ed., *A History of the Canning Industry: Souvenir of the Seventh Annual Convention of the National Canners' Association and Allied Associations* (Baltimore: Canning Trade, 1914), 83.
6. *New York Times*, October 11, 1887, 14; Edward S. Judge, "The Past, Present and Future of the Canned Food Industry," in Arthur I. Judge, ed., *A History of the Canning Industry: Souvenir of the Seventh Annual Convention of the National Canners' Association and Allied Associations* (Baltimore: Canning Trade, 1914), 55; Cox, "The Evolution of Tomato Canning Machinery," 83–84.
7. Cox, "The Evolution of Tomato Canning Machinery," 84–85; Judge, "The Past, Present and Future of the Canned Food Industry," 56.
8. Cox, "The Evolution of Tomato Canning Machinery," 84–85; Judge, "The Past, Present and Future of the Canned Food Industry," 56.
9. "Campbell's Soup," *Fortune* 12 (November 1935):68.
10. *Camden Daily Courier*, September 13, 1913, in the clipping file, and "Campbell Soup Company Historical Synopsis," undated mimeographed paper, both in the Campbell Soup Company archives, Camden, New Jersey; "Campbell's Soup," *Fortune* 12 (November 1935):132.
11. Gordon Morrison, "Tomato Varieties," Special Bulletin no. 290 (East Lansing: Michigan State College, Agricultural Experiment Station, April 1838), 39; Douglas Collins, *America's Favorite Food: The Story of Campbell Soup Company* (New York: Harry N. Abrams, 1994), 131.

12. "Campbell's Soup," *Fortune* 12 (November 1935):132; Morrison, "Tomato Varieties," 43–44, 45.
13. Edward B. Voorhees, "Experiments on Tomatoes," Bulletin no. 63 (New Brunswick: New Jersey Agricultural Experiment Station, 1889); Byron D. Halsted, "Report of the Botanist," *16th Annual Report* (New Brunswick: New Jersey Agricultural Experiment Station, 1895), 293–297; Byron D. Halsted, "Experiments with Tomatoes: Old Varieties Grown the Present Season," *26th Annual Report* (New Brunswick: New Jersey Agricultural Experiment Station, 1904–1905, 1906), 447–477; Carl Raymond Woodward and Ingrid Nelson Waller, *New Jersey's Agricultural Experiment Station, 1880–1930* (New Brunswick: New Jersey Agricultural Experiment Station, 1932), 297.
14. Edward B. Voorhees, "Tomato Growing," Farmers' Bulletin no. 76 (Washington, D.C.: Government Printing Office, 1898).
15. Woodward and Waller, *New Jersey's Agricultural Experiment Station, 1880–1930*, 78, 264–268, 296–305,
16. Woodward and Waller, *New Jersey's Agricultural Experiment Station, 1880–1930*, 299–303.
17. "Campbell's Soup," *Fortune* 12 (November 1935):134; Earl Chapin May, *The Canning Clan: A Pageant of Pioneering Americans* (New York: Macmillan, 1937), 345; Morrison, "Tomato Varieties," 53; Ingrid Nelson Waller, *Where There Is Vision: The New Jersey Agricultural Experiment Station, 1880–1955* (New Brunswick, N.J.: Rutgers University Press, 1955), 53–54.
18. *Landreth's Pennsylvania Certified Tomato Seed* (Bristol, Pennsylvania: D. Landreth Seed Co., [1936]), n.p.; Morrison, "Tomato Varieties," 53.
19. Waller, *Where There Is Vision*, 53.
20. Waller, *Where There Is Vision*, 53–55.
21. Waller, *Where There Is Vision*, 53–55.
22. "Campbell's Soup," *Fortune* 12 (November 1935):132.
23. May, *The Canning Clan*, 345; John T. Cunningham, *Garden State: The Story of Agriculture in New Jersey* (New Brunswick, N.J.: Rutgers University Press, 1955), 70–76.
24. John T. Cunningham, *Garden State: The Story of Agriculture in New Jersey* (New Brunswick: Rutgers University Press, 1955), 76–78.
25. Campbell Soup Company, "The Tomato—From Columbus to Campbell," ca. 1974, n.p.
26. Campbell Soup Company, "The Tomato—From Columbus to Campbell," ca. 1974, n.p.
27. Finbar Kenneally, O.F.M., *Writings of Fermín Francisco de Lasuén, O.F.M.*, 11 vols. (Washington, D.C.: Academy of American Franciscan History, 1965), 2:336; Marcelino Marquinez, letter to Sr. Governor in Santa Cruz, California, dated December 13, 1816, San Francisco Chancery Archives.
28. Edward Douglas Branch, *Westward: The Romance of the American Frontier* (New York: D. Appleton, 1930), 487; Fred A. Shannon, *The Farmer's Last Frontier: Agriculture, 1860–1897*, vol. 5 of *The Economic History of the United States* (New York: Holt, Rinehart and Winston, 1961), 28.

29. *Daily Alta California,* August 14, 1851, 2; *Daily Alta California,* September 13, 1851, 2.
30. *California Farmer* 1 (January 26, 1854):26, 6 (October 10, 1856):93, (December 12, 1856):155, 7 (June 13, 1857):170; *California Culturist* 1 (September 1858):190.
31. Isidor Jacobs, "The Rise and Progress of the Canning Industry in California," in Arthur I. Judge, ed., *A History of the Canning Industry: Souvenir of the Seventh Annual Convention of the National Canners' Association and Allied Associations* (Baltimore: Canning Trade, 1914), 31; Betty Zumwalt, *Ketchup, Pickles, Sauces: 19th-Century Food in Glass* (Fulton, Calif.: Mark West Publishers, 1980), 86; William Braznell, *California's Finest: The History of the Del Monte Corporation and the Del Monte Brand* (San Francisco: Del Monte Corporation, 1982), 15–16, 29–30.
32. *Trade* 24 (June 7, 1901):n.p.; A. McGill, "Lime Juice and Catsup," Bulletin no. 83 (Ottawa: Laboratory of the Inland Revenue Department, November 24, 1902), 18–19; A. McGill, "Tomato Ketchup," Bulletin no. 368 (Ottawa: Laboratory of the Inland Revenue Department, March 31, 1917), 16–17.
33. *American Agriculturist* 28 (August 1869):283.
34. *San Jose Patriot,* as cited in *California Horticultural and Floral Magazine* 1 (February 1871):126.
35. *California Horticultural and Floral Magazine* 1 (August 1871):314, (October 1871):372.
36. Walter Olson, "Changing Patterns of Tomato Production in California," *California Geographer* 10 (1969):16.
37. *American Grocer* 15 (January 23, 1876):117.
38. *Scientific American,* 2nd ser. 44 (April 30, 1881):281.
39. *Semi-Tropical California and Southern California Horticulturist* 4 (November 1881):187.
40. *Semi-Tropical California and Southern California Horticulturist* 4 (November 1881):187; Harris Newmark, *Sixty Years in Southern California (1853–1913)* (New York: Knickerbocker Press, 1926), 428.
41. Kate Sanborn, *A Truthful Woman in Southern California* (New York: D. Appleton, 1893), 128.
42. Edward J. Wickson, *The California Vegetables in Garden and Field* (San Francisco: Pacific Rural Press, 1897), 288; Olson, "Changing Patterns of Tomato Production in California," 15–26.
43. A. I. Dichmont, *Interviews with Persons Involved in the Development of the Mechanical Tomato Harvester* (Davis: Regents of the University of California, 1978); Olson, "Changing Patterns of Tomato Production in California," 16, 24; Charles Plummer, "U.S. Tomato Statistics, 1960–90," Statistical Bulletin no. 841 (Washington, D.C.: USDA, Economic Research Service, August 1992), 14.
44. "Camden Factory Destroyed," *Philadelphia Daily News,* November 4, 1991, 42; Kevin Riordan, "Campbell Demolition Saddens 5 Retirees," *Courier-Post,* November 1, 1991, 4B; Maureen Graham, "A Camden Landmark No Longer," *Philadelphia Inquirer,* November 4, 1991, clipping from the Campbell Soup Company archives, Camden, New Jersey.

45. Hubert G. Schmidt, *Agriculture in New Jersey: A Three-Hundred-Year History* (New Brunswick, N.J.: Rutgers University Press, 1973), 266.
46. Plummer, "U.S. Tomato Statistics, 1960–90," 44.
47. The information in the rest of this section of the chapter is based on interviews with Richard Orzalli, director, Agricultural Operations, Campbell Soup Company, Sacramento, California, and Fran DuVernois, former vice president, Global Operations, Campbell Soup Company, Camden, New Jersey.
48. R. W. Robinson, "A History of the Tomato Genetics Cooperative," *Report of the Tomato Genetics Cooperative* 32 (May 1982):1–2.
49. Charles Rick, "Research Notes," *Report of the Tomato Genetics Cooperative* 5 (January 1955):8–13; Charles Rick, "Linkage Summary," *Report of the Tomato Genetics Cooperative* 27 (February 1977):2–5; Robinson, "A History of the Tomato Genetics Cooperative," 1–2.
50. Charles Rick, "Genetic Resources in Lycopersicon," in Donald J. Nevins and Richard A. Jones, eds., *Tomato Biotechnology* (New York: Alan R. Liss, 1987), 19; Benedict Carey, "Tasty Tomatoes: Now There's a Concept," *Health* 7 (July-August 1993):26; Donald Woutat, "Toward a Tastier Tomato," *Los Angeles Times,* July 7, 1993; John Seabrook, "Tremors in the Hot House," *New Yorker* 69 (July 19, 1993):32–41.
51. *Request for Advisory Opinion: FLAVR SAVR Tomato: Status as Food* ([Davis, California: Calgene, 1991]); G. Kalloo, ed., *Genetic Improvement of Tomato* (New York: Springer-Verlag, c. 1991); Keith Redenbaugh, et al., *Safety Assessment of Genetically Engineered Fruits and Vegetables: A Case Study of the Flavr Savr Tomato* (Boca Raton, Florida: CRC Press, c. 1992); Philip J. Hilts, "Genetically Altered Tomato Moves toward U.S. Approval," *New York Times*, April 9, 1994, 143.

7. AN AMERICAN CULINARY ICON

1. Steve Jones, "Andy Warhol's Allegorical Icons," in Alan R. Pratt, ed., *The Critical Response to Andy Warhol* (Westport, Conn.: Greenwood Press, 1997), 283.
2. Patrick S. Smith, *Andy Warhol's Art and Films* (Ann Arbor: UMI Research Press, 1986), 220; Victor Bockris, *Warhol* (New York: Da Capo Press, 1997), 149.
3. *Time* 79 (May 11, 1962):52; Bockris, *Warhol*, 105.
4. Bockris, *Warhol*, 144–145.
5. Smith, *Andy Warhol's Art and Films*, 219; Bockris, *The Life and Death of Andy Warhol* (New York: Bantam Books, 1989), 109.
6. Smith, *Andy Warhol's Art and Films*, 219–220; Bockris, *Warhol*, 149.
7. *Time* 79 (May 11, 1962):52; Bockris, *Warhol*, 148.
8. Kynaston McShine, ed., *Andy Warhol: A Retrospective* (New York: Museum of Modern Art, 1989), 55.
9. Jones, "Andy Warhol's Allegorical Icons," 281.
10. Bockris, *Warhol*, 149.
11. Ralph Rugoff, "Albino Humour," in Colin MacCabe, ed., *Who Is Andy*

Warhol? ([Dorcester, U.K.]: British Film Institute and Andy Warhol Museum, 1997), 100.

12. Jones, "Andy Warhol's Allegorical Icons," 281.
13. Jones, "Andy Warhol's Allegorical Icons," 281–282.
14. Andy Warhol, as quoted in Bockris, *Warhol*, 144.
15. Donald Judd, "Andy Warhol," *Arts Magazine* 37 (January 1963):49.
16. Marianne Hancock, "Soup's On," *Arts Magazine* 39 (May/June 1965):16–18.
17. Hancock, "Soup's On," 16–18.
18. Jones, "Andy Warhol's Allegorical Icons," 280; George Lois, *Covering the '60s: George Lois, the Esquire Era* (New York: Monacelli, 1996), 6.
19. Bockris, *Warhol*, 160.
20. Lawrence Alloway, "Art," *Nation* 202 (May 24, 1971):668–689.
21. Barbara Goldsmith, "The Philosophy of Andy Warhol," *New York Times Book Review,* September 14, 1975, 3–5.
22. Steven Kurtz, "Uneasy Flirtations: The Critical Reaction to Warhol's Concepts of the Celebrity and of Glamour," in Alan R. Pratt, ed., *The Critical Response to Andy Warhol* (Westport, Conn.: Greenwood Press, 1997), 253.
23. Jones, "Andy Warhol's Allegorical Icons," 281.
24. Letter from W. L. White, assistant counsel, Campbell Soup Company, to Post Originals Ltd., dated January 17, 1967, and letter from Jerald Ordover to W. L. White, dated January 20, 1967. The originals of these letters are in the Andy Warhol Museum, Pittsburgh, Pennsylvania.
25. David and Micki Young, *Campbell's Soup Collectibles: A Price and Identification Guide* (Iola, Wis.: Krause Publications, 1998), 92.
26. Steve Kaufman's Art Studio On-Line: http://www.stevekaufman.com/campbells/campbell_gallery.htm.
27. Young and Young, *Campbell's Soup Collectibles*; phone interview with David Young, January 16, 1999. Those interested in information can write to Soup Collectibles Club, 414 Country Lane, Wauconda, IL 60084.
28. Interview with Patti Campbell, January 20, 1999. For further information about this club, write to CC International, Ltd., 305 East Main St., Ligonier, Pa. 15658 or e-mail her at Member@soupkid.com.
29. Interview with Nancy Bailey, January 20, 1999.
30. Press releases from the Campbell Soup Company: October 19, 1994; May 15, 1995; June 7, 1995; September 5, 1996; March 12, 1997.
31. Eric Asimov, "Homely Soup, Dressed Up and Ready to Go," *New York Times*, November 22, 1996, C1, C26.
32. Laura Yee, "Soup: Aromatic Elixirs," *Restaurants and Institutions* (October 15, 1998), as presented on their Web site, www.rimag.com/820/food-ly.htm.
33. Asimov, "Homely Soup, Dressed Up and Ready to Go," C26.
34. *New York Post*, November 12, 1997, 44.
35. *New York Times,* January 25, 1999, A7.
36. *New York Times,* January 25, 1999, A7.
37. Brigid Allen, *The Soup Book* (London: Macmillan, 1995), 5.
38. *Gourmet* 54 (June 1954):56, (August 1954):27; 55 (March 1955):30; 56 (November 1956):23; 57 (April 1957):27; 58 (June 1958):23; 59 (August

1959):28; 61 (May 1961):38; 63 (July 1963):50; 64 (June 1964):30, 64 (July 1964):21; 65 (December 1965):44; 67 (August 1967):37; 70 (July 1970):80; 70 (August 1970):50.

39. Elizabeth Robins Pennell, *The Feasts of Autolycus: The Diary of a Greedy Woman* (London: John Lane; New York: Merriam, 1896), 211–212.
40. Alice B. Toklas, *The Alice B. Toklas Cookbook* (New York: Harper, 1954), 49–51.
41. James Peterson, *Splendid Soups: Recipes and Master Techniques for Making the World's Best Soups* (New York: Bantam Books, 1993), 207–210.
42. *Mexican Cookery for American Homes* (San Antonio: Gebhardt Chili Powder Company, 1923), 13; Assistance League of Southern California, *The Palatists Book of Cookery* (Hollywood: Citizen-News, 1933), 199, 214; *California Mexican Cook Book* (Los Angeles: E. C. Ortega Company [est. 1934]), n.p.; Erna Fergusson, *Mexican Cookbook* (Albuquerque: University of New Mexico Press, 1945), 22; Diana Kennedy, *The Cuisines of Mexico* (New York: Harper & Row, 1972), 191–192.
43. Early Italian cookbooks published in the United States include Antonia Isola, *Simple Italian Cookery* (New York: Harper & Brothers, 1912); Jack Cusimano, *Economical Italian Cook Book* (Los Angeles: n.p., 1917); Julia Lovejoy Cuniberti, *Practical Italian Recipes for American Kitchens* (Janesville, Wis.: Gazette Printing, 1917). More recent Italian cookbooks are Nancy Verde Barr, *We Called It Macaroni: An American Heritage of Southern Italian Cooking* (New York: Alfred A. Knopf, 1996); Lynne Rossetto Kasper, *The Splendid Table: Recipes from Emilia-Romagna, the Heartland of Northern Italian Food* (New York: William Morrow, 1992); Mary Taylor Simeti, *Pomp and Sustenance: Twenty-five Centuries of Sicilian Food* (Hopewell, N.J.: Ecco Press, 1998); Norman Wasser-Miller, *Soups of Italy: Cooking Over 130 Soups the Italian Way* (New York: William Morrow, 1998), 91–115.
44. *The International Soup Market* (New York: Packaged Facts, 1996), 3, 5, 9.

HISTORICAL RECIPES

1. Recipe for Peanut Cream: "For making cream, the peanuts should not be roasted so much as for making butter. They should have a light straw color. Then grind them very fine, and to a tablespoonful of nut butter add 1½ cups of water, adding a little at a time, and beating until it is done." For Peanut Milk: "Make like the peanut cream, only add more water. The amount of nut butter to be used depends upon the richness of the milk desired." Almeda Lambert, *Guide for Nut Cookery* (Battle Creek, Michigan: Joseph Lambert, 1899), 72.
2. The recipe for boiling peanuts is as follows: "First blanch the peanuts, which can be done by heating in the peanut roaster or in the oven until they are quite hot, but not browned in the least. Let them cool in a dry place, and when nearly cold, the skins can easily be removed by rubbing in the sieve or between the hands, or they can be blanched by pouring boiling water on them, and letting them stand in it until the skins become loosened; then rub off with the hands. When blanched,

put into cold soft water to cook, as they cook quicker in soft water than in hard water—in about one half the time. When perfectly tender, salt and let stew until they are well seasoned throughout. Serve hot." Lambert, *Guide for Nut Cookery*, 251–252.

3. Lambert provides seven recipes for Nutmeatose. The first is as follows: "Take 2 cups of peanut butter, 2½ cups of water, 3 tablespoonfuls of No. 3 gluten, and 1 teaspoonful of salt. Mix all the ingredients together very thoroughly, and cook in cans. (Sealed cans are preferable.) Cook from three to five hours." Lambert, *Guide for Nut Cookery*, 97.

Bibliography and Resources

ON CANNING AND PRESERVING

American Can Company. *Facts about Commercially Canned Foods.* New York: American Can Company, 1936.

______. *Nutritive Aspects of Canned Foods: A Bibliography of Scientific Reports and Helpful Tables of Food Data.* New York: American Can Company, 1937.

______. *The Canned Food Reference Manual.* New York: American Can Company, 1939.

Appert, [Nicholas]. [Trans. by K. G. Bitting.] *The Book for All Households; or, The Art of Preserving Animal and Vegetable Substances for Many Years.* Chicago: n.p., August 1920.

Atwater, W. O., and A. P. Bryant. "The Chemical Composition of American Food Materials." Bulletin no. 28. Washington, D.C.: USDA, 1906.

Benson, O. H. "Home Canning by the One-Period Cold-Pack Method." Farmers' Bulletin no. 839. Washington, D.C.: USDA, June 1917.

Bitting, A. W. *Appertizing; or, The Art of Canning.* San Francisco: Trade Pressroom, 1937.

Bitting, A. W., and K. G. Bitting. *Canning and How to Use Canned Goods.* Washington, D.C.: National Canners Association, 1916.

Blits, H. *Patented and Improved Methods of Canning Fruits with New Edition and Supplement.* Brooklyn: H. I. Blits, 1890.

Caldwell, Joseph S. "Farm and Home Drying of Fruits and Vegetables." Farmers' Bulletin no. 984. Washington, D.C.: USDA, June 1918.

Campbell, Clyde H. *Campbell's Book: A Text Book on Canning, Preserving and Pickling.* [New York: Canning Age, c. 1929.]

______. *Campbell's Book: Canning, Preserving and Pickling.* 2nd ed. New York: Canning Age, 1937.

______. Rev. by Rohland Isker and Walter MacLinn. *Campbell's Book: Canning, Preserving and Pickling.* 3rd ed. Chicago: Vance Publishing, 1950.

Canned Fruits, Vegetables and Food. Collection of twenty U.S. Department of Agriculture and state agricultural bulletins, 1903–1915.

Canning Trade. *A Complete Course in Canning.* 3rd ed. Baltimore: Canning Trade, 1914.

______. *A Complete Course in Canning.* 5th ed. Baltimore: Canning Trade, 1924.

Clark, Hylam. *The Tin Can Book: The Can as Collectible Art, Advertising Art and High Art.* New York: New American Library, Times Mirror, 1977.

Collins, James H. *The Story of Canned Foods.* New York: E. P. Dutton, 1924.

Consumer Guide. *The Food Preserver.* Skokie, Illinois: Publications International, 1976.

Creswell, Mary E., and Ola Powell. "Canning, Preserving, Pickling." *Cooperative Extension Work in Agriculture and Home Economics,* SRS Doc. 22, Ext. 8, no. A-8L. September 1, 1915.

______. "Home Canning of Fruits and Vegetables as Taught to Canning Club

Members in the Southern States." Farmers' Bulletin no. 853. Washington, D.C.: USDA, July 1917.

Cruess, W. V. *Commercial Fruit and Vegetable Products: A Textbook for Student, Investigator and Manufacturer.* 2nd ed. New York and London: McGraw-Hill, 1938.

Culinary Arts Institute. *The Canning and Freezing Book.* Chicago: Consolidated Book Publishers, 1976.

Duckwall, Edward Wiley. *Canning and Preserving of Food Products with Bacteriological Technique,* vol. 1. Pittsburgh: Pittsburgh Printing, 1905.

Folin, Otto. *Preservatives and Other Chemicals in Foods: Their Use and Abuse.* Cambridge: Harvard University Press, 1914.

Harland, Marion [Mary Virginia Terhune]. *The Story of Canning and Recipes.* [Washington, D.C.]: National Canners Association, 1910.

Heath, Ambrose. *Good Dishes from Tinned Foods.* London: Farber and Farber, 1943.

Hoskins, Thomas H. *What We Eat: An Account of the Most Common Adulterations of Food and Drink.* Boston: T.O.H.P. Burnham, 1861.

Judge, Arthur I., ed. *A History of the Canning Industry: Souvenir of the Seventh Annual Convention of the National Canners' Association and Allied Associations.* Baltimore: Canning Trade, 1914.

Keuchel, Edward F., Jr. *The Development of the Canning Industry in New York State to 1960.* Ph. D. dissertation, Cornell University, 1970.

Lake View Woman's Club. *Balanced Meals with Recipes: Food Values, Drying and Cold Pack Canning, Menus with and without Meat, Box Luncheons.* Chicago: Lake View Woman's Club, 1917.

Lee, John A. *Canned Foods: How to Buy; How to Sell.* Baltimore: Canning Trade, 1914.

Lemcke, Gesine. *Preserving and Pickling.* New York: D. Appleton, 1899.

Linton, F. B. *Inspection of Fruit and Vegetable Canneries.* USDA Bulletin no. 1084. Washington, D.C.: Government Printing Office, 1922.

Loomis, Henry M. *The Canning of Foods and Some Tested Recipes.* Bulletin no. 100-A. Washington, D.C.: National Canners Association, 1929.

May, Earl Chapin. *The Canning Clan: A Pageant of Pioneering Americans.* New York: Macmillan, 1937.

Meyer, Hazel. *Hazel Meyer's Freezer Book.* Philadelphia and New York: J. B. Lippincott, 1970.

Mrs. Price's Complete Directions for Canning Vegetables and Fruits, Pickling, etc. Minneapolis: Price Compound, n.d.

National Canners Association. *The Canning Industry.* Washington, D.C.: National Canners Association, March 1, 1939.

______. *The ABC's of Canned Foods.* Rev. ed. Washington, D.C.: National Canners Association, 1961.

Neil, Marion H. *Canning, Preserving and Pickling.* Philadelphia: David McKay, 1914.

Pacrette, Jean. *The Art of Canning and Preserving as an Industry.* New York: Henry I. Cain & Son, 1901.

Parloa, Maria. "Canned Fruit, Preserves, and Jellies: Household Methods of

Preparation (Corrected March 25, 1905)." Farmers' Bulletin no. 203. Washington, D.C.: USDA, 1905.

______. *Canned Fruit, Preserves and Jellies: Household Methods of Preparation.* Chicago: Saalfield Publishing, 1917.

Peterson, M. E. *Preserving, Pickling and Canning Fruit.* Philadelphia: G. Peterson, 1869.

Porter, Virginia, and Esther Latzke. *The Canned Foods Cook Book.* New York: Doubleday, Doran, 1939.

Powell, Ola. *Lippincott's Home Manuals: Successful Canning and Preserving.* Philadelphia: J. B. Lippincott, 1917.

Schwaab, Ernest. *The Secrets of Canning: A Complete Exposition of the Theory and Art of the Canning Industry.* Baltimore: John Murphy, 1890.

Sim, Mary B. *Commercial Canning in New Jersey: History and Early Development.* Trenton: New Jersey Agricultural Society, 1951.

Taylor, Demetria. *The Complete Book of Home Canning.* New York: Greenburg, 1943.

Tinned Foods and How to Use Them. London: Ward, Lock & Bowden, 1893.

Wright, Mary M. *Preserving and Pickling.* Philadelphia: Penn Publishing, 1924.

SELECTED BIBLIOGRAPHY ON SOUP

Berolzheimer, Ruth, ed. *250 Delicious Soups.* Chicago: Culinary Arts Institute, 1940.

Binsted, Raymond, and James D. Devey. *Soup Manufacture, Canning, Dehydration and Quick-Freezing.* 2nd ed., rev. and enlarged. London: Food Trade Press, 1960.

Brown, Cora, Rose Brown, and Bob Brown. *Soups, Sauces, and Gravies.* Philadelphia and New York: J. B. Lippincott, 1939.

Cameron, Miss. *Soups and Stews and Choice Ragouts.* London: Kegan Paul, Trench, Trübner, 1890.

The Campbell Museum Collection. 2nd ed. Camden, New Jersey: Campbell Museum, 1972.

Campbell Soup Company. *Wonderful Ways with Soups.* Camden, New Jersey: Campbell Soup Company, 1958.

______. *A Campbell's Cookbook: Cooking with Soup.* [Camden, New Jersey]: Campbell Soup Company, 1962.

______. *Campbell's Great Restaurants Cookbook, U.S.A.* [New York: Rutledge Books, 1969?].

______. *A Campbell Cookbook: Easy Ways to Delicious Meals.* Rev. ed. [Camden, New Jersey]: Campbell Soup Company, 1970.

______. *Campbell's 100 Best Recipes plus 157 Other Family Favorites and Party Dishes.* Camden, New Jersey: Campbell Soup Company, ca. 1974.

______. *Most-for-the-Money Main Dishes.* [Camden, New Jersey: Campbell Soup Company], c. 1975.

______. *A Campbell Cookbook: Most-for-the-Money Main Dishes.* [Camden, New Jersey]: Campbell Soup Company, 1977.

______. *Campbell's Great American Cookbook.* New York: Random House, c. 1984.

______. *Campbell's Creative Cooking with Soup*. New York: Beekman House, 1985.

______. *Campbell's Great American Cookbook*. New York: Weathervane Books, distr. by Crown Publishers, 1989.

______. *Pace Family Recipe Round-up*. Alexandria, Virginia: Time-Life Books; Camden, New Jersey: Campbell Soup Company, c. 1996.

______. *Healthy Cooking Made Easy*. Des Moines, Iowa: Meredith Custom Publishing, c. 1997.

Castle, Coralie. *Soup*. Santa Rosa, California: 101 Productions/Cole Group, 1992.

Carey, Nancy. *Soup to Nuts*. Philadelphia: Smith, 1929.

Collins, Douglas. *America's Favorite Food: The Story of Campbell Soup Company*. New York: Harry N. Abrams, 1994.

De Gouy, Louis P. *The Soup Book: 770 Recipes*. New York: Greenberg, 1949.

Elliot, Virginia, and Robert Jones. *Soups and Sauces*. New York: Harcourt, Brace, 1934.

The Everyday Soup Book: A Recipe for Every Day of the Year including February 29th by G. P. London: Stanley Paul, 1915.

Ewing, Emma P. *Soups and Soup Making*. Chicago and New York: Fairbanks, Palmer, 1882.

For Variety Cook with Soup. Pittsburgh: H. J. Heinz Company, 1977.

Franklin, Rena. *Soups of Hakafri Restaurant*. Kosher ed. Drawings [by] Daniel; trans. [by] Yehudit Venezia. Gainesville, Florida: Triad, c. 1981.

Green, Olive. *One Thousand Simple Soups*. New York: G. Putnam's Sons, 1907.

Guest, [Flora Bigelow]. *Soups, Oysters and Surprises*. London: John Lane, 1918.

Heath, Ambrose. *Good Soups*. London: Farber & Farber, 1935.

Herrick, Christine Terhune, and Marion Harland. *Consolidated Library of Modern Cooking and Household Recipes. Book III: Soups, Chowders, and Fish; Meats, Poultry, and Game*. New York: R. J. Bodmer, 1904.

Hoff, Archie Croydon. *Soups and Consommés*. Los Angeles: International Publishing, 1913.

Howe, Robin. *Soups*. New York: Bonanza Books, 1967.

"Huckins' Soups." Wholesale price list and advertisements. Boston: J.H.W. Huckins & Co., n.d.

The International Soup Market. New York: Packaged Facts, 1996.

Joseph Campbell Company. *Campbell's Menu Book*. Camden, New Jersey: Joseph Campbell Company, 1910.

______. *Campbell's Condensed Tomato Soup*. Camden, New Jersey: Joseph Campbell Company, 1914.

______. *Helps for the Hostess*. Camden, New Jersey: Joseph Campbell Company, 1916.

Larousse, David Paul. *The Soup Bible*. New York: John Wiley & Sons, 1997.

Lederman, Martin. *The Slim Gourmet's Soup Book*. New York: Dutton, 1958.

Mabon, Mary Frost. *A Meal in Itself: A Book of Soups*. New York: Eagle Books, distr. by Duell, Sloan and Pearce, 1944.

Murrey, Thomas J. *Fifty Soups*. New York: White, Stokes & Allen, 1884.

Northcross, Frances Troy. *250 Delicious Soups*. Chicago: Culinary Arts Institute, 1940.

Peterson, James. *Splendid Soups: Recipes and Master Techniques for Making the World's Best Soups*. New York: Bantam Books, 1993.

Rose, Evelyn. "Replicating the Taste of Home Made Soup in a Canned Product." In Tom Jaine, ed., *Oxford Symposium on Food and Cookery 1987: Taste, Proceedings*. London: Prospect Books, 1988, 180–182.

Ruf, Fritz, ed.*"Die sehr bekannte dienliche Löffelspeise" Mus, Brei und Suppe—kulturgeschichtlich betrachtet.* Velbert-Neviges, Germany: BreRing Verlag, 1989.

Standard, Stella. *Stella Standard's Soup Book*. New York: Taplinger, 1978.

Vachon, Joseph. *Book of Economical Soups and Entrées*. Chicago: Hotel Monthly Press, 1903.

Wasser-Miller, Norman. *Soups of Italy: Cooking Over 130 Soups the Italian Way*. New York: William Morrow, 1998.

Whitehead, Jessup. *Hotel Meat Cooking*. 7th ed. Chicago: Jessup Whitehead, 1901.

Young, David, and Micki Young. *Campbell's Soup Collectibles: A Price and Identification Guide*. Iola, Wisconsin: Krause Publications, 1998.

Your Soup Menu. [Pittsburgh: H. J. Heinz Company], 1995.

SELECTED BIBLIOGRAPHY ON THE TOMATO

American Horticultural Society. *Illustrated Encyclopedia of Gardening Tomatoes*. Mount Vernon, Virginia: American Horticultural Society, 1982.

Behr, Edward. "A Flavorful Tomato, Revisited." *Art of Eating* 9 (Winter 1989):1, 7.

Calgene, Inc. *Request for Advisory Opinion: FLAVR SAVR Tomato: Status as Food*. [Davis, California: Calgene, 1991]

Corbett, Wilfred. "The History of the Tomato." In *Fifteenth Annual Report of the Experimental and Research Station*. London: Cheshunt, 1930, 82–83.

Dillon, Clarissa. "Tomato Mania." *Living History* 1 (Summer 1991):1–8.

Fawcett, W. Peyton. "Happiness Is a Ripe Love Apple." *Field Museum of Natural History Bulletin* 41 (August 1970):2–5.

Gould, Wilbur. *Tomato Production, Processing and Quality Evaluation*. Westport, Connecticut: AVI Publishing, 1974.

______. *Tomato Production, Processing and Quality Evaluation*. 2nd ed. Westport, Connecticut: AVI Publishing, 1983.

______. *Tomato Production, Processing and Technology*. 3rd ed. Baltimore: CTI Publications, 1992.

Grewe, Rudolf. "The Arrival of the Tomato in Spain and Italy: Early Recipes." *Journal of Gastronomy* 3 (Summer 1987):67–83.

Hester, Jackson Boiling. *Soil Fertility in Tomato Production*. [Riverton, New Jersey]: Campbell Soup Company, Department of Agricultural Research, c. 1940.

Hester, Jackson Boiling, and Florence A. Shelton. *The Soil Side of Tomato Growing*. [Riverton, New Jersey]: Campbell Soup Company, Department of Agricultural Research, 1939.

Jenkins, J. A. "The Origin of the Cultivated Tomato." *Economic Botany* 2 (October-December 1948):379–392.

Kalloo, G., ed. *Genetic Improvement of Tomato*. New York: Springer-Verlag, ca. 1991.

Luckwill, Leonard C. The Genus *Lycopersicon*: An Historical, Biological and Taxonomic Survey of the Wild and Cultivated Tomatoes. Aberdeen University Studies no. 120. Aberdeen: University Press, 1943.

______. "The Evolution of the Cultivated Tomato." *Journal of the Royal Horticultural Society* 68 (1943): 19–25.

McCue, George A. "The History of the Use of the Tomato: An Annotated Bibliography." *Annals of the Missouri Botanical Garden* 39 (November 1952):289–348.

Moore, John Adam. "The Early History of the Tomato or Love Apple." *Missouri Botanical Garden Bulletin* 23 (October 1935):134–138.

Muller, Cornelius. *A Revision of the Genus Lycopersicon*. Washington, D.C.: USDA, Miscellaneous Publication no. 382, 1940.

______. "The Taxonomy and Distribution of the Genus Lycopersicon." *National Horticultural Magazine* 19 (July 1940):157–160.

Olson, Walter. "Changing Patterns of Tomato Production in California." *California Geographer* 10 (1969):15–26.

Redenbaugh, Keith, et al. *Safety Assessment of Genetically Engineered Fruits and Vegetables: A Case Study of the Flavr Savr Tomato*. Boca Raton, Florida: CRC Press, c. 1992.

Report of the Tomato Genetics Cooperative, 1951–present.

Richman, Irwin. "The History of the Tomato in America." *Proceedings of the New Jersey Historical Society* 80 (July 1962):151–173.

Shepherd, Steven. *In Praise of Tomatoes: A Year in the Life of a Home Grower*. New York: HarperCollins, 1996.

Smith, Andrew F. "Tomato Pills Will Cure All Your Ills." *Pharmacy in History* 33 (1991):169–180.

______. "The Great Tomato Pill War of the Late 1830s." *Connecticut Historical Society Bulletin* 56 (Winter-Spring 1991):91–107.

______. "The History of Home-made Anglo-American Tomato Ketchup." *Petits Propos Culinaires* 39 (December 1991):35–45.

______. "Dr. John Cook Bennett's Tomato Campaign." *Old Northwest* 16 (Spring 1992):61–75.

______. "Authentic Fried Green Tomatoes?" *Food History News* 4 (Summer 1992):1–2.

______. "The Amazing Archibald Miles and His Miracle Pills: Dr. Miles' Compound Extract of Tomato." *Queen City Heritage* 50 (Summer 1992):36–48.

______. *Bibliography on the Tomato in America to 1860: Annotated and Indexed*. New York: Tomato History and Culture Project, 1993.

______. "Canning and Bottling Tomatoes in Nineteenth-Century America." *Food History News* 6 (Summer 1994):1–2, 6.

______. "The Early History of the Tomato in Florida." In C. S. Vavrina, ed., *Proceedings of the Florida Tomato Institute*. [Immokalee]: University of Florida, Horticultural Sciences Department, Institute of Food and Agricultural Sciences, 1994, 123–133.

______. "Early American Tomato Varieties." *Off the Vine* 1 (1994):1, 6.

______. "The Early History of the Tomato in South Carolina." *Carologue* 11 (Summer 1995):8–11.

______. *The Tomato in America: Early History, Culture and Cookery.* Columbia: University of South Carolina Press, 1994.

______. *Pure Ketchup: The History of America's National Condiment.* Columbia: University of South Carolina Press, 1996.

______. "The Culinary Historian as Detective: The Case of Who Really Ate the First Tomato in America." *International Cookbook Revue* 1, no. 5 (1996):73–74.

______. "The Tomato Pill War and the Introduction of the Tomato into New England." In Peter Benes, ed., *The Dublin Seminar for New England Folklife Annual Proceedings 1995: Plants and People.* Boston: Boston University, 1996.

______. "From Garum to Ketchup: A Spicy Tale of Two Fishy Sauces." In Harlan Walker, ed., *Fish: Food from the Waters; Proceedings of the Oxford Symposium on Food and Cookery, 1997.* Devon, U.K.: Prospect Books, 1998.

Weber, George F. "A Brief History of Tomato Production in Florida." *Proceedings of the Florida Academy of Sciences* 4 (1940):167–174.

TOMATO COOKBOOKS

Bailey, Lee. *Tomatoes.* New York: Clarkson Potter, 1992.

Bay Books. *The Amazing Tomato Cookbook.* Kensington, New South Wales: Bay Books, n.d.

Behr, Edward. "A Flavorful Tomato, Revisited." *Art of Eating* 9 (Winter 1989):1, 7.

Bevona, Don. *The Love Apple Cookbook.* New York: Funk and Wagnalls, 1968.

Brennan, Ethel, and Georgeanne Brennan. *Sun-Dried Tomatoes.* San Francisco: Chronicle Books, 1995.

Carver, George Washington. "How to Grow the Tomato and 115 Ways to Prepare It for the Table." Bulletin no. 36. Tuskegee, Alabama: Tuskegee Institute Experiment Station, 1936.

Cool, Jesse Ziff. *Tomatoes: A Country Garden Cookbook.* San Francisco: Collins Publishers, 1994.

Croce, Julia Della. *Salsa di Pomodoro: Making Great Tomato Sauces in Italy.* San Francisco: Chronicle Books, 1996.

Dribin, Lois, Denise Marina, and Susan Ivankovich. *Cooking with Sun-Dried Tomatoes.* Tucson: Fisher Books, 1990.

DuBose, Fred. *The Total Tomato: America's Backyard Experts Reveal the Pleasures of Growing Tomatoes at Home.* New York: Harper and Row, 1985.

Garden Way Publishing. *Tomatoes! 365 Healthy Recipes for Year-Round Enjoyment.* Pownal, Vermont: Storey Communications, 1991.

Guste, Roy F. *The Tomato Cookbook.* Gretna, Louisiana: Pelican Publishing, 1995.

Hendrickson, Robert. *The Great American Tomato Book: The One Complete Guide to Growing and Using Tomatoes Everywhere.* Garden City, New York: Doubleday, 1977.

Hill, Nicola. *The Tomato Cookbook.* Philadelphia: Courage Books, 1995.

Hobson, Phyllis. *Great Green Tomato Recipes.* Bulletin A-24. Pownal, Vermont: Storey Communications, 1978.

Hoffman, Mable. *The Complete Tomato Cookbook*. New York: HP Books, 1994.

Jordan, Michele Anna. *The Good Cook's Book of Tomatoes, with More Than 200 Recipes*. Reading, Massachusetts: Addison-Wesley, 1995.

Michaelson, Mike. *The Great Tomato Cookbook*. Chicago: Greatlakes Living Press, 1975.

Nimtz, Sharon, and Ruth Cousineau. *Tomato Imperative!* New York: Little, Brown, 1994.

Old-Fashioned Tomato Recipes, including Green Tomato Recipes. Nashville, Indiana: Bear Wallow Books, 1981.

Simmons, Paula. *The Green Tomato Cookbook*. Seattle: Pacific Search, 1975.

Siegel, Helene. *The Totally Tomato Cookbook*. Berkeley: Celestial Arts, 1996.

Turman, Marianne, and Ray Turman. *Cooking with Fresh Tomatoes*. Sun City Center, Florida: Printed for the author, 1992.

Weir, Joanne. *You Say Tomato*. New York: Broadway Books, 1998.

SELECTED WORKS ON TOMATO CULTURE

Blancard, Dominique. *A Colour Atlas of Tomato Diseases*. London: Manson Publishing; New York: Halsted Press, 1995.

Day, J. W. *A Treatise on Tomato Culture*. Crystal Springs, Mississippi: By the author, 1891.

Doty, Walter. *All about Tomatoes*. San Francisco: Ortho Books, 1981.

Foster, Catherine. *Terrific Tomatoes*. Emmaus, Pennsylvania: Rodale, 1975.

Haber, Ernest Straign. *The Influence of the Soil Reaction on the Ionizable Constituents of the Tomato as Determined by Electrodialysis*. [Ames, Iowa: n.p., 1929.]

Krech, Inez M. *Tomatoes*. New York: Crown, 1981.

Livingston, A. W. *Livingston and the Tomato with a Preface and Appendix by Andrew F. Smith*. Reprint. Columbus: Ohio University Press, 1998.

Luebbermann, Mimi. *Terrific Tomatoes: Simple Secrets for Serious Gardens, Indoors and Out*. San Francisco: Chronicle Books, 1994.

National Gardening Association. *Tomatoes: Growing, Cooking, Preserving*. New York: Villard, 1987.

Page, John. *Grow the Best Tomatoes*. Bulletin A-27. Pownal, Vermont: Storey Communications, 1979.

Pellett, Frank C., and Melvin A. Pellett. *Practical Tomato Culture*. New York: A. T. De La Mare, 1930.

Raymond, Dick, and Jan Raymond. *The Gardens for All Book of Tomatoes*. Burlington, Vermont: National Association for Gardening, 1983.

Root, A. I., J. W. Day, and D. Cummins. *Tomato Culture*. Medina, Ohio: A. I. Root, 1892.

______. *Tomato Culture*. 2nd ed. Medina, Ohio: A. I. Root, 1906.

Rundell, Mary G. *Texas Gardener's Guide to Growing Tomatoes*. Waco, Texas: Suntex Communications, 1984.

Rupp, Rebecca. *Blue Corn and Square Tomatoes: Unusual Facts about Common Garden Vegetables*. Pownal, Vermont: Garden Way, 1987.

Salisbury, J. H. *Account of Some Investigations Connected with the Composition of the Tomato*. [Albany: n.p., 1848.]

Smith, F. F. *Tomatoes from Seed to the Table.* Aurora, Illinois: Knickerbocker & Hodder, 1876.

Shepherd, Steven. *In Praise of Tomatoes.* New York: HarperCollins, 1996.

Tarr, Yvonne Young. *The Tomato Book.* New York: Vintage, 1976.

Tomato Facts: The Story of the Evolution of the Tomato. Columbus, Ohio: Livingston Seed Company, 1909.

Tracy, William W. *The Tomato Culture.* New York: Orange Judd, 1907.

Watterson, John C. *Tomato Diseases: A Practical Guide for Seedsmen, Growers & Agricultural Advisors.* Saticoy, California: Petroseed Company, 1985.

Wittwer, S. H., and S. Honma. *Greenhouse Tomatoes, Lettuce and Cucumbers.* East Lansing: Michigan State University Press, 1979.

Work, Paul. *The Tomato.* New York: Orange Judd, 1942.

SELECTED TOMATO BULLETINS AND GOVERNMENTAL REPORTS

Bailey, Liberty Hyde. "Notes on Tomatoes." Bulletin no. 19. Lansing: Agricultural College of Michigan, 1886.

______. "Notes on Tomatoes." Bulletin no. 31. Lansing: Agricultural College of Michigan, 1887.

Beattie, James H. "Greenhouse Tomatoes." Farmers' Bulletin no. 1431. Washington, D.C.: USDA, 1924, rev. 1935.

Boswell, Victor R. "Improvement and Genetics of Tomatoes, Peppers, and Eggplant." In *Yearbook, 1937.* Washington, D.C.: USDA, 1938.

Breazeale, J. F. "Canned Tomatoes, Catchup, Chow-Chow, etc." In "Canning Tomatoes at Home and in Club Work." Farmers' Bulletin no. 521. Washington, D.C.: USDA., 1913.

Carncross, John W., et al. "Putting Profit in Tomato Growing." Circular no. 409. New Brunswick: New Jersey Agricultural Experiment Station, April 1941.

Carver, George Washington. "How to Grow the Tomato and 115 Ways to Prepare It for the Table." Bulletin no. 36. Tuskegee, Alabama: Tuskegee Institute Experiment Station, 1936.

Cook, Mel T. "Diseases of Tomatoes." Circular no. 104. New Brunswick: New Jersey Agricultural Experiment Station, 1918.

Coons, G. H., and Ezra Levin. "The Leaf-Spot Disease of Tomato." Special Bulletin no. 81. East Lansing: Michigan Agricultural College Experiment Station, 1917.

Corbett, L. C. "Tomato Growing as Club Work in the North and West." Washington, D.C.: USDA, Bureau of Plant Industry, February 20, 1915.

Courter, J. W., and J. S. Vandemark. "Growing Tomatoes at Home." Circular no. 981, rev. Urbana: University of Illinois, College of Agriculture, Cooperative Extension Service, July 1971.

DeBaun, R. W. "Early-Tomato Growing in New Jersey." Circular no. 103. New Brunswick: New Jersey Agricultural Experiment Station, 1919.

Doolittle, S. P. "Tomato Diseases." Farmers' Bulletin no. 1934, rev. Washington, D.C.: USDA, March 1948.

Dutt, J. O. "Growing Tomatoes in the Home Garden." Leaflet no. 124. University Park: Pennsylvania State University, February 1948.

Dutt, J. O., and R. F. Fletcher. "Growing Tomatoes in the Home Garden." Circular no. 477. University Park: Pennsylvania State University, n.d.

Fletcher, S. W., and O. I. Gregg. "Pollination of Forced Tomatoes." Special Bulletin no. 39. East Lansing: Michigan State College Experiment Station, October 1907.

Fromme, F. D. "Experiments in Spraying and Dusting Tomatoes." Bulletin no. 230. Blacksburg: Virginia Polytechnic Institute, Virginia Agricultural Experiment Station, November 1922.

Fromme, F. D., and H. E. Thomas. "Spraying and Dusting Tomatoes." Bulletin no. 213. Blacksburg: Virginia Polytechnic Institute, Virginia Agricultural Experiment Station, December 1916.

Gardner, Max W., James B. Kendrick, and L. C. Cochran. "Dusting Tests for the Control of Tomato Diseases, 1926." *Transactions Indiana Horticultural Society* (1927):75–83.

Halsted, Byron D. "Experiments with Tomatoes: Old Varieties Grown the Present Season." In *26th Annual Report.* New Brunswick: New Jersey State Agricultural Experiment Station, 1904–1905, 1906. Pp. 447–477.

Howard, Burton J. "The Sanitary Control of Tomato-Canning Factories." Bulletin no. 569. Washington, D.C.: USDA, June 25, 1917.

Huelsen, W. A. "Wilt-Resistant Tomato Varieties Released by the Illinois Station." Circular no. 490. Urbana: University of Illinois, Agricultural Experiment Station, January 1939.

Joffe, Jacob Samuel. "Suggestions to Tomato Growers." Circular no. 391. New Brunswick: New Jersey Agricultural Experiment Station, 1939.

Kreutzer, W. A., and L. W. Durrell. "Collar Rot of Tomatoes." Bulletin no. 402. Fort Collins: Colorado Agricultural College, Colorado Experiment Station, August 1933.

Lesley, J. W., John T. Middleton, and C. D. McCarty. "Simi, a Processing Tomato Resistant to Verticillium and Fusarium Wilts." *Hilgardia* 21 (February 1952):289–299.

McWhorter, F. P., and J. A. Milbrath. "The Tipblight Disease of Tomato." Station Circular no. 128. Corvallis: Oregon State College, Agricultural Experiment Station, August 1938.

Meyers, C. E. "The Pennheart Tomato." Bulletin no. 438. State College: Pennsylvania State College School of Agriculture and Experiment Station, January 1943.

Meyers, C. E., and M. T. Lewis. "The Effect of Selection in the Tomato." Bulletin no. 248. State College: Pennsylvania State College School of Agriculture and Experiment Station, January 1930.

Minges, P. A., et al. "Tomato Propagation." Circular no. 160. Berkeley: University of California, January 1950.

Morrison, Gordon. "Tomato Varieties." Special Bulletin no. 290. East Lansing: Michigan State College, Agricultural Experiment Station, April 1938.

O'Dell, Charles R. "Production of Caged Tomatoes." Publication no. 420. Blacksburg: Extension Division, Virginia Polytechnic Institute and State University, June 1972.

Plummer, Charles. "U.S. Tomato Statistics, 1960–90." Statistical Bulletin no. 841. Washington, D.C.: USDA, Economic Research Service, August 1992.

Porte, William S. "The Pritchard Tomato." Circular no. 243. Washington, D.C.: USDA, September 1932.

______. "The Pan America Tomato, a New Red Variety Highly Resistant to Fusarium Wilt." Circular no. 611. Washington, D.C.: USDA, June 1941.

______. "Development of Interspecific Tomato Hybrids of Horticultural Value and Highly Resistant to Fusarium Wilt." Circular no. 584. Washington, D.C.: USDA, January 1941.

______. "Commercial Production of Greenhouse Tomatoes." Farmers' Bulletin no. 2082. Washington, D.C.: USDA, May 1955.

Pritchard, Fred J. "Development of Wilt-Resistant Tomatoes." Bulletin no. 1015. Washington, D.C.: USDA, Bureau of Plant Industry, March 28, 1922.

Pritchard, Fred J., and William S. Porte. "The Control of Tomato Leaf-Spot." Bulletin no. 1288. Washington, D.C.: USDA, December 19, 1924.

______. "The Break o' Day Tomato." Circular no. 218. Washington, D.C.: USDA, March 1932.

Schroeder, W. T. "Control of Tomato Diseases by Spraying." Bulletin no. 724. Geneva: New York State Agricultural Experiment Station, Cornell University, June 1947.

Scott, F. H. "Fruit-Set Hormones for Earlier Tomatoes." Circular no. 523. Blacksburg: Extension Division, Virginia Polytechnic Institute and State University, n.d.

Shapovalov, Michael. "Experiments on the Control of Tomato Yellows." Technical Bulletin no. 189. Washington, D.C.: USDA, July 1930.

Stair, Edw. C. "Indiana Baltimore Tomato—Its History and Development." Circular no. 207. Lafayette, Indiana: Purdue University Agricultural Experiment Station, December 1934.

Sturtevant, E. Lewis. "The Tomato." Special Bulletin. College Station: Maryland Agricultural Experiment Station, 1889. Pp. 18–25.

"Studies on Tomato Leaf-Spot Control." Bulletin no. 345. New Brunswick: New Jersey Agricultural Experiment Station, November 1920.

Tiedjens, V. A., and O. W. Davidson. "Tomatoes in the Greenhouse." Circular no. 443. New Brunswick: New Jersey Agricultural Experiment Station, Rutgers University, June 1942.

Tiedjens, V. A., et al. "Growing Tomatoes in New Jersey Home Gardens." Bulletin no. 21, number 11. New Brunswick: College of Agriculture and Agricultural Experiment Station, Rutgers University, May 1944.

"Tomato Diseases and Insect Pests: Identification and Control." Circular no. 809. Urbana: University of Illinois, July 1959.

"Tomato Diseases in Ohio." Bulletin no. 321. Wooster: Ohio Agricultural Experiment Station, February 1918.

Voorhees, Edward B. "Experiments on Tomatoes." Bulletin no. 63. New Brunswick: New Jersey Agricultural Experiment Station, 1889.

______. "Tomato Growing." Farmers' Bulletin no. 76. Washington, D.C.: Government Printing Office, 1898.

Wellington, Richard. "Influence of Crossing in Increasing the Yield of the Tomato." Bulletin no. 346. Geneva: New York Agricultural Experiment Station, March 1912, 57–76.

BIBLIOGRAPHY AND RESOURCES

ON NEW JERSEY AND THE TOMATO

Bridgeton Evening News, December 11, 1936; May 18, 1936 [250th Anniversary Supplement].

Casper, Charles W. Untitled report on canning in Salem County. Unpublished manuscript at the Salem County Historical Society, dated March 13, 1906.

Cunningham, John T. *Garden State: The Story of Agriculture in New Jersey*. New Brunswick: Rutgers University Press, 1955.

______. "Love-Apple of Yesteryear." *Newark Sunday News* (magazine section), February 6, 1955, 19–22.

Heston, Alfred M., ed. *South Jersey: A History, 1664–1924*. New York: Lewis Historical Publishing, 1924.

Holbrook, Stewart H. *Lost Men of American History*. London: Macmillan, 1946.

Moore, George. "Farm Record 1853–77." Unpublished handwritten manuscript from Cumberland County, in the Rutgers University Library (Special Collections), New Brunswick, New Jersey.

Pitt, T. Dimitry, and Lewis P. Hoagland. *New Jersey Agriculture: Historical Facts and Figures*. Trenton: New Jersey Department of Agriculture, 1943.

Richman, Irwin. "The History of the Tomato in America." *Proceedings of the New Jersey Historical Society* 80 (July 1962):151–173.

Schmidt, Hubert G. *Rural Hunterdon: An Agricultural History*. New Brunswick: Rutgers University Press, 1945.

______. *Agriculture in New Jersey: A Three-Hundred-Year History*. New Brunswick: Rutgers University Press, 1973.

Sim, Mary. *Commercial Canning in New Jersey: History and Early Development*. Trenton: New Jersey Agricultural Society, 1951.

Smith, Andrew F. "Robert Gibbon Johnson and the Tomato." *New Jersey History* 108 (Fall/Winter 1990):59–74.

Ten Ton Tomato Club of New Jersey *Bulletin*, 1940–1954.

Waller, Ingrid Nelson. *Where There Is Vision: The New Jersey Agricultural Experiment Station, 1880–1955*. New Brunswick: Rutgers University Press, 1955.

Weygandt, Cornelius. *Down Jersey: Folks and Their Jobs, Pine Barrens, Salt Marsh, and Sea Islands*. New York: D. Appleton-Century, 1940.

Woodward, Carl Raymond. *The Development of Agriculture in New Jersey, 1640–1880*. New Brunswick: New Jersey Agricultural Experiment Station, 1927.

______. *Ploughs and Politicks: Charles Read of New Jersey and His Notes on Agriculture, 1715–1774*. New Brunswick: Rutgers University Press, 1941.

Woodward, Carl Raymond, and Ingrid Nelson Waller. *New Jersey's Agricultural Experiment Station, 1880–1930*. New Brunswick: New Jersey Agricultural Experiment Station, 1932.

Wright, William C., and Paul A. Stellhorn. *Directory of New Jersey Newspapers, 1765–1970*. Trenton: New Jersey Historical Commission, 1977.

NEW JERSEY COOKBOOKS AND WORKS ON CULINARY HISTORY

Bishop, Anne. *The Victorian Seaside Cookbook*. Newark: New Jersey Historical Society, c. 1983.

Christ Church. *Tried and True Recipes: The Home Cook Book*. East Orange, New Jersey: Christ Church, 1889.

Economical Cookery. Newark, New Jersey: Benjamin Olds, 1839, 1840.

First Women's Guild. *Up to Date*. Montclair, New Jersey: Baptist Church, 1909.

400 Recipes or Recipes of "The 400." A Compilation of Choice Recipes of the Wives, Sisters and Friends of the Members of the Union Club, of Rutherford, N.J. Rutherford: Bergen County Herald Office, [1892].

Galsgie, Eduard Carl. *Diet for Health with Favorite Health Home Recipes*. Jamesburg, New Jersey: n.p., 1913.

Gillis, Mary M. *Food Efficiency; or, The Best Food for the Least Money*. Jersey City, New Jersey: International Letter Club, [1920].

Hatrak, Amy, comp. *Fanny Pierson Crane: Her Receipts, 1796; Confections, Savouries, and Drams. Being a Collection of Sixty Favourites, and Including a Proper Collation and Supper Menu. A Guide to Open Hearth and Bee Hive Oven Cookery, Prepared in the 18th Century Manner, and Adapted for Twentieth Century Use*. Montclair, New Jersey: Montclair Historical Society, 1972.

Howell, Sarah Biddle. *Nine Family Dinners and How to Prepare Them*. [Trenton, New Jersey: Naar, Day & Naar, 1890.]

Johnson's New Family Receipt Book. Boonsford, New Jersey: Odd Fellows Office, 1852.

Keen, Georgia Harmony, comp. *Some Famous Old Recipes*. Elizabeth, New Jersey: Printed for the author by Journal Press, 1904.

Kirmess Cook Book: A Collection of Well-Tested Recipes from the Best Housekeepers of Jersey City and Elsewhere. Compiled for the Kirmess Given for the Benefit of Christ Hospital of Jersey City, Nov. 7-8-9-10, 1906. [Jersey City? ca. 1907.]

Ladies of the North Reformed Church. *Cooks in Clover: Reliable Recipes*. [Passaic, New Jersey]: Thurston & Barker, 1889.

Ludlum, L. M. *The New and the Old in Cookery*. Paterson, New Jersey: Craig, Beckmeyer, 1891.

Methodist Episcopal Church. *The Indispensable Cook Book*. Plainfield, New Jersey: Ladies Committee, n.d.

Nichols, William H. *The Science and Art of Good Cookery*. [n.p.: n.p.,] 1912.

Pompton Plains, New Jersey, First Dutch Reformed Church. *First Dutch (Reformed) Cook Book*. [New York: Polhemus,] 1883.

[Rolfe, Mrs. John Henry, comp.] *Receipt Book*. New Brunswick, New Jersey: Improvement Society of the Second Reformed Church, c. 1890.

Swain, Rachel. *Cooking for Health; or, Plain Cookery with Health Hints*. 3rd ed. Passaic, New Jersey: Health-Culture Company, 1909.

Second Reformed Church, Young People's Association. *Table Talk: A Collection of Tried and Approved Recipes*. [Hackensack, New Jersey: Democrat, 1886.]

Tenafly Presbyterian Church, Ladies Aid Society. *The Palisades Cook Book*. Tenafly, New Jersey: Ladies' Aid Society, [ca. 1910].

Thornton, P. *Southern Gardener*. Newark, New Jersey: A. L. Dennis, 1845.

Townsend, Gore. *The Star Cook Book*. Jersey City, New Jersey: Star Publishing, 1895.

The Woodbridge Cook Book. Woodbridge, New Jersey: Ladies' Association of the First Congregational Church, 1903.

Young Women's Christian Temperance Union. *Y's Cook-book for Wise Cooks.* 3rd ed. Salem, New Jersey: Young Women's Christian Temperance Union, 1893.

Index

INDEX

ABOUT THE AUTHOR

Andrew F. Smith is an independent scholar and freelance writer. He teachers culinary history at the New School University in Manhattan and serves as a consultant to television series on food history. He is the author of *The Tomato in America: Early History, Culture and Cookery* (University of South Carolina Press, 1994), *Pure Ketchup: The History of America's National Condiment* (University of South Carolina, Press, 1996) and *Popped Culture: A Social History of Popcorn in America* (University of South Carolina Press, 1999).